Aston Martin
Engine Development

1984–2000

Old Pond PUBLISHING

Aston Martin
Engine Development

1984–2000

Arthur G. Wilson

First published 2015

Published by
5M Publishing Ltd,
Benchmark House,
8 Smithy Wood Drive,
Sheffield, S35 1QN, UK
Tel: +44 (0) 1234 81 81 80
www.5mpublishing.com

A catalogue record for this book is
available from the British Library

ISBN 978-1-910456-08-8

Book layout by Mark Paterson
Printed by Replika Press Pvt. Ltd., India

Photos by Arthur G Wilson

Arthur G Wilson

Arthur George Wilson was born in Bletchley, Bucks in June 1936. He was educated at the local secondary school and then at Wolverton Technical College. He started a six year apprenticeship at the London Brick Company and specialised in engines. He joined Aston Martin in the early part of 1959 on engine build and soon transferred to the rectification department to look after engines and engine tuning. He then moved to the experimental department as test bed engineer. This progressed to development engineer and senior development engineer at Aston Martin Tickford. During this time he worked on the early Lagonda, the Vantage and the 1980 versions of all three engines and many of the racing customer's engines including the John Pope Special twin turbo, the Tickford endurance race engine and the V12 Weslake engine, the Tickford Capri and the Frazer Metro.

In 1984 he was offered the position in charge of engine development back at AML where he developed the 580X Vantage engine, the 32-valve engine and the design and development for the twin supercharged 32-valve Vantage engine, and the engine for the normally aspirated V8 Coupe. He is married with one daughter and has an interest in classic cars, which has resulted in him restoring an MGA and an MG Midget. He also owns a Jaguar XJ6 and a 1988 Vauxhall 3.0 GSi Carlton.

In Memory of my friend
John Pope who passed away
peacefully 18th June 2013.

Contents

Foreword by David Morgan

I was very honoured when Arthur asked me to write a forward for his book. We are both engineers at heart, and had very similar introductions to life inside Aston Martin. In my case I joined Aston in 1964, in a position as a fitter in the Experimental Department. It was an excellent time to join Aston, as the Experimental Department was just being transferred to Newport Pagnell and the existing staff didn't want to leave London! Initially we commuted up to Feltham daily, usually in the 'Brown Bomber', a rare early Aston fitted with an estate body and a DB4 engine!

Once settled in the three-storey building at Newport Pagnell, I was fortunate to be handed the build of the first V8 and continued to be involved with the V8 until I left Aston, so as you can imagine I have great affection for it! I continued to be fortunate to rise within Aston Martin, to engineer and eventually chief development engineer. As you might expect there were many high and low points during this period. Some of the memorable ones are developing and supporting the Lola Aston Martin at Le Mans in 1967. While unsuccessful at Le Mans, the exercise resulted in a major redesign and the very strong V8 engine that went into production.

During the early 1970s, I had been promoted to senior development engineer and I needed someone to replace my hands-on work and Arthur was the perfect choice. He was working his magic on the production cars at the end of the line and I needed those skills in engineering. I was soon to realise that I had a genius on my team.

1

Arthur's skill and enthusiasm is very rare, he has a great love of cars and of engines in particular, as you will discover while reading his book. While Arthur mainly worked on engine projects, the Bosch Injection systems and subsequent carburettor versions, this was Aston Martin and you had to be flexible! Arthur would be involved in all of the many projects – from driving emission durability cars and riding shotgun in the Bulldog to running all manner of engines on the dynamometer. Whatever Arthur was involved in he gave it 100 per cent and his attention to detail was always impressive. Arthur's restored cars are perfect examples of his exceptional abilities and his eye for detail, they are all absolutely immaculate.

Among the lighter moments were the trips to Le Mans to support the V8 engines in the various chassis – the long 24-hour vigilance; the hectic pit stops to identify misfired and block off injectors on the non-firing cylinders but still managing to get the car home in a well-earned seventh place; the 'if onlys', like when a clutch cylinder leak lost one of the cars a top five finish. We had some happy times too!

While Aston Martin has been well documented over the years, this book gives you a unique detailed inside look into a period of Aston's history from a person who was a part of actually making it happen.

I write this foreword in the grateful appreciation of Arthur's friendship and help to me over very many years. I'm full of admiration at the abilities he so clearly demonstrated in the Aston Martin 32 valve Vantage that he created.

David Morgan.
Formerly Chief development engineer, Aston Martin.
(1964-1981) Tickford, Engineering Ltd.
Director of Engines (1981–2002).

Author's Preface

W hen I first started to write these notes I was not sure what I would do with them or why I should write about the period of AML engine history that happened during my stewardship. But I had started at a time when so much of what had gone before appeared to be of little value and was being thrown away. I suppose that I just felt that with all of the changes that were occurring during the mid-nineties there was a strong possibility that an important part of our history would be lost forever or worse still, rewritten. I am by nature a shy person, much happier in a test bed or on a test track with a box of jets and E tubes rather than in any spotlight. I have been interviewed by journalists and authors on a number of occasions and I can only say that I sympathise with them – the term "blood out of a stone" springs to mind. So if I don't actually write something down myself then my version of events will surely be lost. Therefore I am a slightly reluctant author and I fear not a very polished one – my first attempt and all that but here goes.

I have written about engine development at AML mainly during the period 1984 to 2000 but with some background reference to my time at AML leading up to that period. I write with the reasonably well-informed Aston Martin enthusiast in mind, sufficiently well informed to understand the effects of legislative constraints on motor manufacture as opposed to the relative freedom of aftermarket modification and racing. Naturally my comments in this book will mainly apply to the former. It is simply a book about engine development at AML.

One thing that became very obvious as I wrote was the pace at which technology progresses within the motor industry, particularly the components used to respond to environmental issues. When I started to write I was writing about the technical innovation of recent times but as time went on I was aware of being overtaken by the pace of change. Technical innovation is a very perishable subject to write about so the content in this book on that subject should be looked at with the early to mid-90s in mind. What may have seemed an almost insurmountable problem at that time and therefore a major achievement when resolved would perhaps now seem a minor inconvenience due to the advances made in component technology, particularly as regards catalytic converters and electronic engine management. Having said that, anyone who knows the business will appreciate that the 32 valve Vantage engine in particular did set new standards. It was the most powerful production car engine at the time of its release and the design philosophy behind the supercharged approach was appropriate for the challenges of the time.

London Brick Co to
Aston Martin Lagonda

I was about seven years of age when I first fell in love with a motor car. Up to that time and bearing in mind that I was born in the same year as the Spitfire made its first flight, it had been aircraft that had been my main interest. The subject of my new affection was our first family car, a 1934 Wolseley Hornet Special known as Phoebe II. As was the practice of the time, the name had been professionally sign-written on the offside front of the long bonnet and was there when my parents bought the car. To a young lad, Phoebe was everything that a real sports car should be. The engine was a lovely little six-cylinder with single overhead cam, twin Skinners Union (SU) carburettors and loads of chrome oil pipes. It made a wonderful noise and smelled of hot oil mixed with a sort of hot electrical insulation material, all scary stuff. It still gives me a tingle when I think about it, funny how smell is such an important part of one's memories of old cars. From then on I was hooked, particularly on 1930s sports cars.

When I left school at the age of 15, I began a six-year apprenticeship as a motor fitter at the London Brick Company (LBC) at Newton Longville, note the old term 'fitter' rather than 'mechanic'. I would be working on the fleet of lorries used to transport the company product, not the sports or racing cars that I would have liked. But I was told that the standard of fitting required on diesel engines of the time was much higher than for the run-of-the-mill petrol engine and would stand me in good stead for later. Also the LBC provided one of the best apprentice schemes local to me at that time.

The lorries were mostly diesel engine AEC with a couple of ERF that I remember did not have starter motors. They were started on the handle that in the depths of winter sometimes required a team of men on a long rope and a guy in the cab with a flare down the intake to get them started. The flare was a piece of welding wire twisted around a piece of rag at the end, which was dunked into the fuel tank, set alight and held over the air intake while the engine was cranked. The flare procedure was common practice for all of the diesel engines in cold weather. There was also a number of side valve petrol engine Morris Commercials for more local deliveries and a small fleet of vans, buses, coaches and an ambulance to maintain.

We occasionally had manager's company cars in for work. These were mostly Morrises or Wolseleys, which I remember as increasing in Royal Automobile Club-rated horsepower in line with the status of the manager. The chairman of the company had a beautiful dark green Aston Martin DB2/4, which sometimes came in for a wash and polish. Unfortunately that was all – we didn't get to work on the car; I expect that it went back to Feltham for any service work. But it was a rather special treat for those of us who were motor enthusiasts just to be able to admire such a beautiful car at close quarters. He also had an ash green Jaguar XK120 that came in for the same treatment. I must admit that I slightly favoured the looks of the XK120; it was just so stunningly beautiful with a great looking engine. But I also recognised that underneath that beautiful body, the mechanical bits were less impressive – the workmanship on the Aston was far superior. The engineering on the Aston had a beauty of its own and it was really just as good looking but in a more refined way.

I completed my apprenticeship in June 1957, having specialised on engines. At the beginning of 1959 I was surprised to see an advert for engine builders at the Aston

Martin factory at Newport Pagnell. I knew that Aston Martin had a factory at Feltham and hadn't realised that the company had set up just a few miles away from Bletchley where I lived. With the image of that beautiful DB2/4 in mind I just had to apply. So after making an appointment by phone and armed with every qualification, certificate, commendation and prize certificate that I could lay my hands on I drove to Newport Pagnell in my pride and joy, an Austin Healey 100.

It would be an understatement to say that I was very nervous during the interview; even the slightest possibility of working for such a prestigious motor manufacturer gave me goose pimples. I just had to get the job. Happily I did and started soon after as third man in what was then the DB4 engine pre-production build area in the Olympia building, so called because of the shape of the roof. This was before the end was cut off at an angle for road widening, which if I remember correctly happened during the winter of 1960. I remember that the pay was slightly less than I was getting at the LBC and I would have to drive to work each day. But I didn't mind too much, after all I was going to be working for Aston Martin.

I have one regret about that time and that is that in the excitement I didn't have the grace to properly thank the LBC guys that I had been working with and in particular the manager Chris Johnson for all of their help and patience during my apprenticeship. Self-centred youth, I suppose.

Starting at Aston Martin Lagonda

Starting at Aston Martin Lagonda (AML) engine build, Ray Perkins had the first bench, George Wardle had the second and I had the next one. For the first week I was put to work with George to get to know what was expected. He introduced me to the subtleties of building an aluminium engine as opposed to the iron lumps that I was used to. Starting on the same day but at about 11 o'clock was my long-term friend Les Brooker.

Behind me there were a number of old hands who were building the engines for the DB2/4 Mk III, which was the current production car of the time. They were Fred Woodland, Harry Absolem, Seth Costall and Fred Watters. Eric Pointer made up the subassemblies and prepared the engines for installation into the chassis after the engine test. Sid King balanced crankshafts, etc. Fred Woodland was later to become famous as Mr Duckhams in the engine oil adverts of the time. They were a friendly lot and full of fun. I recall that there was a slightly mischievous attempt at winding up rivalry between Harry Absolem, who had a TR2, and the new boy, me with my Austin Healey. But it was all in good fun.

One of the highlights of my first year at AML was when all of the drivers for the 1959 Le Mans team visited the engine shop prior to the race.

All engines were run in and power tested by Bob (whose surname I cannot recall) and Ernie Mod. Every fifth engine was then stripped for inspection after power testing. At that time the cylinder heads were built on a separate line by Bob Clare, Peter Riley and Dick Joyce. There was someone else but I am afraid that I can't remember his name; he left shortly after I started. The other

Figure 2.1 Old 1959 photo of me (on the right) with Martin Irish and Marina Jones in the car park before the service department was built. Note the building in the background, which became known as the Sunnyside office. It was then home to the factory manager and had a nice orchard to the left. My Austin Healy is in the background.

(I eventually crashed the Healey on the way to work one morning. I ran into a bus carrying young ladies to the Rodex clothing factory in Bletchley, which is how I met my wife to be, she was on the bus.)

half of the building was used to assemble the Mk III chassis prior to pushing it across Tickford Street to the coach-building side of the factory for body and trim, etc. In those days, the separation of the company into two parts by Tickford Street was a little more than geographic. The Olympia side of the road was Associated Engineering Union territory. The other side was traditionally the Coach Builders Union side, in line with its Tickford coach building roots.

My first introduction to the ups and downs of the exotic motor business was at Christmas 1960. Suddenly there were cutbacks that resulted in people being made redundant. Sadly this included my brother Stuart, who had started that year on inspection in the engine shop. But

I just survived – I would have been next to go if there had been one more redundancy. My friend Les Brooker, who started the same day as me but at 11 o'clock, was made redundant, that's how close it was. *(First in last out was the rule on these unhappy occasions. And last out, first back on the return trip, unless one had blotted one's copy book.)* That was the first of many ups and downs during my time at AML, the closing down at the end of 1974 being the most traumatic for everyone.

About midway through my first year at AML it was discovered that there was a problem with the connecting rod bolts, which meant that all engines had to be stripped and rebuilt with new bolts. Seth Costal and I were given this task. Soon after we had completed the rebuilds, I developed dermatitis on my hands, possibly due to prolonged exposure to engine oil during this process. This was strange, as I had not had any problems while working on the diesel engines in my previous employment. The outcome of this was that I was put onto building sub-assemblies for a while, but this also made me available to resolve engine-related problems that occurred during engine tests and road tests. Somewhere during my misspent youth I had developed something of a talent for setting up an engine and its carburettors. This was probably due to involving myself in fiddling with most types of multi-carburettor layouts on the various sports or sporting saloons that my friends and in particular my car-mad family had owned. This eventually resulted in a permanent transfer to the road test and rectification department, looking after general engine problem-solving and engine tuning. And then eventually a position in the experimental department of the time and finally to end up in charge of the engine department at Newport Pagnell. So dermatitis is not such a bad thing after all.

Pip Aires and Dick Bolton were the road testers in the road test and rectification department when I worked there, looking after engine tune and rectification. Bill Jackson had been on road test but had moved over to become a service engineer. Just for interest, there is more than a 95 per cent chance that any reader with a six-cylinder car from DB5 onwards will have had his or her engine tuned by me before it left the factory, nearer to 100 per cent if it was a Vantage. The same would apply to the Bosch-equipped V8 engines and early V8 carburettor cars.

Setting up the throttle linkage to get the correct progression on the Bosch V8 can be a bit tricky and as I had developed a procedure for doing this, I will *very* briefly outline the procedure for the benefit of any owners reading this book. Other readers can just skip this part.

Assuming that all else is as per the instruction book for these cars, you will need a threaded 2BA rod with a ball connector to enable connection to the main throttle lever on the cross shaft and a sliding hook over the rod to be hooked over the bonnet landing rubber at the bottom of the scuttle. A knurled 2BA nut is then required to screw onto the rod to enable fine adjustment of the throttle opening to the various settings.

The cross shaft has a lever at each end to operate the throttle opening. These levers have a clamp bolt to enable the lever to be set at zero with throttles closed. The throttle levers are connected to the main throttle assembly by double ended, L&R threaded rods with lock nuts.

Initially I would concentrate on three positions, zero, 4.6 and 50.5 degrees on the metering unit. Zero, 2.5 and 40 degrees throttle respectively. If these positions are correct then all other check positions should fall into place.

Ensure that all pointers read zero and that throttles are closed on the stops with no free play and also that the metering unit is on its closed stop. Adjust the throttle setting tool via the knurled nut to open the throttle to 4.6 degrees metering unit and 2.5 degrees throttle. If this is not correct, recheck the zero setting and also that all three pointers start to move at the same time. Continue to open the throttle to give 50.5 degrees on the metering unit. The throttle indicator pointer should show that both throttles are 40 degrees open. If this is not so, release the lock nuts and adjust the double-ended (left- and right-hand thread) throttle rods, each side to double the error (i.e., if the reading is 42 degrees then double this to read 44 degrees, or if 38 degrees adjust to 36 degrees, and retighten the lock nuts). Adjust the tool to return the metering unit to zero on its closed stop and release the clamp bolt on the outer throttle levers and reset the throttles to zero on the stops, with no free play.

Repeat the above until all three readings are correct, after which a full range check should confirm that all is in order.

Metering unit degrees	Throttles degrees (A & B bank)
Zero	Zero
4.6	2.5
8.8	5.0
12.7	7.5
16.3	10.0
23.1	15.0
29.3	20.0
40.5	30.0
50.5	40.0
59.4	50.0
67.4	60.0
74.4	70.0
81.0	80.7

Figure 2.2 Me in a DB5 Vantage Show car. I can't remember which show but car is left-hand drive with air-conditioning, so could be for the USA.

Figure 2.3 Engine bay of same show car, just been set up.

The DB5 Vantage pictured in Figures 2.2 and 2.3 almost didn't make it to the show. I forget why, but for some reason I had to do a cylinder head change on this car in the rectification department. I had changed the head and had gone off to return the valve timing jig, etc. to the engine build shop as they needed it back. I still had to set the fuel levels and ignition timing before tuning the carbs. While I was away, someone convinced the apprentice of the time, Paul Mintoft, to move the car off of the ramp. There was a spit back in the air box causing a small fire, no big drama until someone decided to take the air box lid off to look at the fire – big mistake. The fire got out of hand and required the use of an extinguisher that made a mess. I ended up doing another head change and the apprentice got a telling off, rather unfairly, I thought, as he had only done as he was told.

Figure 2.4 Silverstone 1959 – a few of us had skived off for the afternoon practise on Friday. I remember two of us standing at the Woodcote end of the pit lane chatting to Stirling Moss, as you could in those days, when an Aston came around Woodcote well out of shape. I think it could well have been driven by Roy Salvadori. Mr Moss seemed to disappear in a puff of smoke only to reappear on the other side of the pit wall; we did our best to follow as the car flashed past, narrowly missing the pit wall.

The road test department was closed down for a short time in July of 1971. During that time I worked at our main London agents as road tester. This was something of a new experience for me and gave me the opportunity of direct contact with customers, which gave me a greater insight into what was important to them. I also developed a great deal of respect for the work that our dealerships and in particular our own service department did. After a little short of a year, engine tune was becoming a problem back at the factory and I was invited back.

As a young apprentice during the 1950s I had followed the fortunes of Aston Martin sports car racing and I am still very much in awe of the people that made Aston Martin what it was in 1959 when I joined the company. That year has to be the pinnacle of what was a long period of outstanding motor racing achievement with nothing remotely like it since. It is a milestone year in Aston Martin history and for that matter in British motor racing history. So even after 41 years at AML, I still feel something of an upstart, especially when contemplating writing anything about the company.

I feel very fortunate to have worked at AML during the period that I did. Starting at the beginning of the year that we won the sports car championship and the Le Mans 24-hour race was a good starting point to a career that has never been dull. It would be difficult to imagine working anywhere else in the motor industry that could provide the same level of involvement in a very skill-dependent process. Maybe a process from a different age where craftsmanship was more necessary and had a greater value than it does nowadays. There were times when we had to work very hard indeed but somehow it didn't seem to matter it was just something that needed to be done. Working at AML was more of a way of life than a job.

Looking back, I suppose that the David Brown years must be considered as the high point in post-war Aston Martin and Lagonda history, a time of strong investment in motor sport to promote the product. When I started work at Newport Pagnell, I started as an employee of the automotive division of David Brown Industries, but as from the close of business on 30 June 1960, this became Aston Martin Lagonda Ltd. Thereafter AML had to stand alone and make its own way as a separate company, which

15

is when life became tougher for all of us. But the magic in the names 'Aston Martin' and 'Lagonda' somehow got us through. They have a well-earned reputation as being something special, even exotic, by motor enthusiasts in all walks of life, including some of our component suppliers who so often supported us with much more enthusiasm than they really needed. This is an image that has encouraged motor enthusiast entrepreneurs to invest their time and fortunes and employees to put a lot more than just doing a job into what they were producing.

There was always closeness and a feeling of belonging to something special working at AML, a strong family relationship that extended out to envelop owners and dealerships worldwide. Working at AML, one always had a great sense of being part of a family, one strong enough to ride the ups and downs of the motor industry while producing very special motor cars for a few very lucky people.

Experimental Department

I n 1972, I was offered a position in the experimental department on engine development, primarily on engine test bed development and carburetion. I would be working for David Morgan, who was chief engineer on engine development reporting to Harold Beech, who was head of design. You can imagine how pleased I was, although it did mean a drop in earnings and any overtime was to be unpaid but expected.

This was a very exciting time for me working directly for David in the experimental department at AML and later on within Aston Martin Tickford (AMT). Also during the Tickford days I was fortunate to work closely with Alistair Lyle, who was chief engine designer for AMT. I learned a great deal from Alistair in the process. They were busy times and it would not be possible to cover it all in this book, but the following is a taster.

The early part of my time in the experimental department was mostly taken up with the move to carburettors on the V8 engine and, in particular, the USA version.

Contrary to some of the more fanciful stories that I have read, the reasons for moving on from the Bosch mechanical injection was no more than the normal process of product development by both parties. Bosch was going away from mechanical injection and we would have been the only people using it (if Bosch would have been prepared to produce and support it for such a small volume). The complete system was also quite heavy and it was quite complex and time-consuming to manufacture and set up when compared to a carburettor installation. Carburettor technology of the time was not standing still and it was possible to achieve a good balance of power and emission control using four Weber twin-choke 42DCNF carburettors. Drive ability was improved, performance was at least on a par and fuel consumption was slightly improved – the only downside was the need for a power bulge in the bonnet. From an engine development point of view, it was much easier to make changes to the carburettor calibration in-house, whereas any changes to the calibration of the Bosch metering unit meant Bosch involvement. Again the question of continued Bosch support has to be considered, taking into account that they were going away from the mechanical system and our small-volume manufacture.

Times were hard for the UK motor industry as a whole during 1973/4, and just before Christmas 1974 we were all taken to a meeting at the local cinema and told that the company was going into receivership and that we would be made redundant. This was a bit of a shock to say the least. Somehow I had to earn a crust, so I started up working for myself rebuilding race engines and tuning carburettors, etc. on various exotics including Astons. I was doing pretty well but when, in 1975, the call came to go back to AML, I couldn't resist – it just gets into your blood. When I started back in what had been the production rectification area there were two other guys already there, Bill Harris and John White. Roger Date re-joined a week after me. We were putting together cars that were part finished and I was required primarily to set the engines up but we all pretty well tackled anything on the car. I still had jobs in progress at my workshop, so for a while I worked all day at AML and at my workshop in the evenings.

Eventually things began to pick up and AML was reborn as AML (1975). And as departments slowly started to regroup I was invited to re-join the experimental department by Mike Loasby, who was now in charge of the department. Harold Beach was retained as a consultant.

Thus began a very busy period as the go-ahead for the new William Towns-styled Lagonda had been agreed. There was a huge enthusiasm and spirit within the revived AML, which was just as well because the target for the new car was set for Earls Court, October 1976. Every effort was put into that car, with everyone doing whatever was needed, whenever it was it needed, I even did coach lining. We made it, just. I think that I fitted the last bit on the stand at Earls Court, the solid silver motif atop the grill. There were some other memorable Lagonda moments as well. Dennis Flather was able to make special arrangements for a Lagonda to enter the 1977 London to Brighton run and I was given the privilege. I believe that this was the first time a modern car had been allowed to enter.

The Lagonda was a completely new car and chassis with a very low bonnet line and a new engine bay. The new engine bay layout would allow us to lower the engine but would require a smaller oil sump capacity for ground clearance and a new oil pump. But even then, the low bonnet line would still restrict the air intake system. To offset this, the compression ratio was raised and larger 2.1" inlet valves were fitted along with new camshafts to boost mid-range torque. Much work was done on the intake system to optimise the limited space available above the carburettor intakes. Various attempts at smoothing the flow by shallow trumpets and different air entry points to ensure even distribution resulted in a gauze dome or flame trap over the carburettor intake being the best, combined with a side-entry air box. The gauze domes really work, particularly in the mid-range. Top-end power was always compromised by the low bonnet line up to the introduction of the 1985 injection engine cars, which used the same engine as the V8 Coupe of the time, so had comparable power.

A higher performance version of the V8 engine was to be developed for a new, more performance-biased version of the V8 Coupe, a return of the Vantage name last used on a six-cylinder car. The engine was developed using 48 IDF carburettors feeding through 1.5" inlet ports to 2.1" diameter inlet valves, new pistons and camshafts and higher compression ratio. A twin four-into-one manifold and exhaust system was developed specific to this engine. It produced 370 bhp at 6,000 rpm with 350 ft/lb of torque at 5,000 rpm. During the road and track development stage we had a real problem with keeping the spark plugs clean on this engine specification, although the heat range was correct for full power running. MSD ignition helped but the eventual cure was to use NGK BP6EV spark plugs, which worked very well. Thus began a long and happy association with NGK and I have always specified NGK plugs ever since.

The car was also extensively changed aerodynamically, with front air dam, blanked radiator grill, Perspex covers on headlamps and a flip tail rear spoiler. All of which resulted in a remarkably low drag factor for such a large frontal area car. This had involved considerable night-time testing in the wind tunnel by the engineers responsible. Night time testing was cheaper. We also did track testing with short lengths of wool attached to the front and side bodywork. I can remember driving the development Vantage down the timing straight at MIRA at an indicated 170 mph, with guys at the side of the track taking photos. The suspension was developed after extensive track and road work to include Koni shock absorbers and Pirelli CN12 tyres on 7" rims.

In those days we had no shortage of good analytical test drivers within the department. We also used the services of people like Ray Mallock and Bill Nicholson. I once had the pleasure of going testing with Bill Nicholson. For some reason I had been off work unwell. I was still feeling a little second-hand when I returned to work, only to be told 'you are going brake fade testing with Bill, that will sort you out'. Brake fade testing is not pleasant for the co-pilot at the best of times, with repeated hard braking from 100 mph upwards. I survived but I have to say that I have never been driven so relentlessly hard for every inch of road between Newport Pagnell and MIRA. Compared to the trip to and from MIRA, the brake fade testing was a breeze, Bill could really drive a car. Bill also did a lot of the durability and chassis development driving on the Lagonda. If it was going to break, Bill would break it. I remember that he knocked off a number of engine oil filters while landing the other side of a particular humpback bridge on the test route before we tucked it up far enough out of harm's way.

The first customer V8 Vantage cars were converted from standard V8s from the production line. They were passed over to experimental to be converted individually in a small area of the experimental workshop. I was made responsible for this process and the sign-off after testing. From memory, I think that the rolling road sign-off required something of the order of 300/310 bhp at the rear wheels. All cars had to achieve an indicated 170 mph on level ground. The cars were then returned to the production line for final finish and valet. One very keen customer had his car delivered straight to Silverstone circuit to enter it in an Aston Martin Owners' Club (AMOC) race, which he won.

USA certification was looming again and eventually I had to go out to the EPA facility in Ann Arbour, Michigan with David Orchard, who was in charge of the safety and legislation department at AML, to certify the USA specification car. This was my first trip abroad for the company but luckily David knew his way around, having been there many times before. There were a few hiccups, not the least of which was the shipping company at

Detroit airport losing track of the container with our certification car in it. But they eventually found it some days after it had arrived. After we had successfully certified the car and had our fill of Ann Arbour's buzzing nightlife, we were invited to stay with Peter Sprague for a few days so we drove upstate, stopping off at the Niagara Falls on the way just to have a look. We enjoyed Peter's hospitality in what is a stunningly beautiful part of America. Over the years I have had many enjoyable trips to America on company business.

DP1054

Engine development was ongoing for the 1980 engine range. At this time we had three different cylinder head specifications: the standard V8 with the 1.9" diameter inlet valves, the higher compression Lagonda head with 2.1" inlet vales and the Vantage with the larger valves and the inlet ports bored out to 1.5" diameter. All had slightly different compression ratios. So it would make sense to commonise engine parts where possible.

Using the Lagonda cylinder head as the basis of the standard 1980 type would mean that all engines used the larger inlet valves and that only minor changes to build procedure and a change of camshafts would be required to produce a Vantage version. The Vantage would need new camshafts to optimise the power from the smaller inlet port and to further refine the engine. John Lipman was a camshaft consultant at that time and was commissioned to design new polynomial camshafts for development work. Further refinement to limit piston noise during warm-up would require a new common piston design with barrel skirt rather than the existing taper skirt. Support for the piston side thrust at bottom dead centre could also be extended 0.100" down the bore by reducing the chamfer at the bottom of the liner. With the new piston design we

22

were able to reduce the clearance in the bore from 0.0045" to 0.0035".

Dished valve heads would be part of development, as they had shown a significant reduction in hydrocarbon emissions during earlier development of the federal specification engine due to greater conformity to the valve seat. Tufrided valve stems would reduce wear particularly with the unleaded fuel required by the USA specification cars.

Fuel consumption was improved by almost 2 mpg during the official test cycle, partly by quite a tricky modified ignition system giving part throttle advance and partly due to an improved auto transmission. The ignition system was only tricky to set up in that it used a double acting vacuum capsule for retard at idle and advance during part throttle cruise. The fact that our engine did not make enough vacuum at part throttle to advance the ignition meant that we used the air pump pressure to assist the advance side by pushing from the retard side of the capsule. This required that the advance be switched out at manifold pressure above 1" hg to avoid damage to the engine.

The *Motor* magazine road test of a Vantage in April 1981 obtained figures of 5.2 seconds to 60 mph and 11.9 seconds to 100 mph; they wrote that it was the fastest production car they had tested up to that date.[*] Comparisons were made with the Porsche 3.3 Turbo at 12.3 seconds to 100 mph and the Lamborghini Countach at 13.1 seconds to 100 mph. They also rated it as the best handling car.

[*] *Ref.* Motor *Road Test No 21/81.*

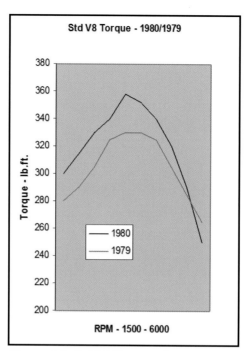

Figure 3.1 Std V8 Torque Sheet 2

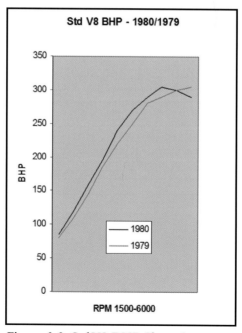

Figure 3.2 Std V8 BHP Sheet 1

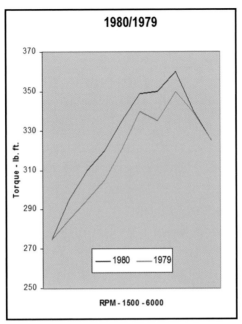

Figure 3.3 Vantage Torque Sheet 1.

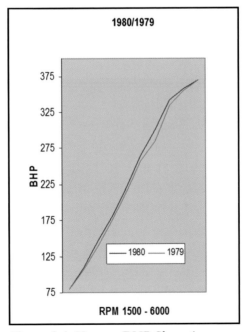

Figure 3.4 Vantage BHP Sheet 1.

25

V12 Engine

DP1080

Another interesting project was the development of the Ford Weslake race engine. In 1979, Aston Martin bought the rights, drawings and all tooling for the Ford Weslake V12 race engine. The thought at the time was that it could be developed as a four-litre engine for a new, smaller Aston Martin.

The engine had originally been designed as a 48-valve, three-litre V12 with 75 mm bore and 56.46 mm stroke, with the potential to stretch to four litres for sports car racing by opening the bore out to 80 mm and increasing the stroke to 66.2 mm. It should not be confused with the earlier Gurney Weslake Formula One engine, although the original engine design was based around the crank and main bearing arrangement for that engine. But the Ford Weslake engine has the benefit of having the five intermediate main bearing caps cross-bolted into the cylinder block for added stiffness.

The first prototype of this engine ran at the end of 1971 and in 1973 it is reputed that two engines were used for evaluation by JW Automotive in the Gulf Mirage endurance race car with favourable results. It was reputed to have been producing about 460 bhp at 10,500 rpm.

The AML development project number given to this engine project was DP1080. There was one complete engine, No. 006, included in our purchase and in April 1979 this was stripped to enable our chief engine designer, Alistair Lyle, to evaluate the design. The engine was then rebuilt in preparation for test bed development work. Our first run was 10 March 1980 and development continued on to the end of June. Initially we limited rpm to 10,000 but towards the end we ran to 11,000 rpm on occasions. We couldn't equal the claimed output for this engine, at least not as fully stabilised readings. But it was an interesting project working on an out and out racing engine, and created some wonderful noises in the process. But at the end of the day it was decided that the cost of transforming it into a road car engine would be similar if not more expensive than starting from a clean sheet of paper. Also the stretch to four litres would be very questionable for a production road car.

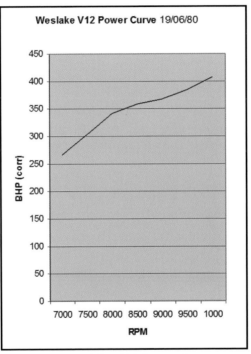

Figure 4.1 Weslake V12 Power Curve.

The main features of the engine were:

- 60 degrees V12.
- Four gear-driven overhead camshafts.
- Lucas Mk II mechanical injection and Lucas RITA ignition.
- Slide throttles.

- Firing order 1-12-5-8-3-10-6-7-2-11-4-9, using the traditional AML method of cylinder numbering.
- 75 mm bore and 56.5 mm stroke giving a swept volume of 2996 cc.
- Crankshaft made from En40B.
- 60.4 mm main bearing.
- 41.24 mm big end journal, 120 degree throw.
- Pent roof combustion chamber.
- Inlet ports were angled at 45 degrees from cylinder axis.
- It was said that a 12/1 compression ratio had been achieved but from memory, the engine that we had was found to be around 11/1.
- The piston crown was extensively machined to provide clearance for valve movement and to maximise the compression ratio.
- Cylinder centres were 93.65 mm apart, allowing for a bore increase to 80mm. This combined with a longer stroke of 66.2 mm would give a swept volume of 3993.6 cc.
- The cylinder block and the cylinder heads were cast in LM25WP, other castings were in Magnesium.
- Dry sump lubrication system was via four oil pumps: two for crankcase scavenge, one for cylinder head scavenge and one for pressure feed to the engine.
- Weight was approximately 385lb: it was just under 32" long, 19" wide and 22" high.
- Maximum safe engine rpm was 11,500 rpm.

Aston Martin Tickford

T he creation of Aston Martin Tickford (AMT) in the early 1980s had provided a means to offset vehicle engineering development and certification costs to AML by selling our engineering expertise to other manufacturers as AMT, as Porsche and Lotus had done. It also helped to finance a much larger engineering department, with the top specialist engineers and the high-tech equipment required to compete in the modern engineering age. That arrangement had worked well initially, with a number of high-profile projects for major motor manufacturers, and could so easily have been the solution that AML needed. But unfortunately, after a few good years, it was decided that AMT should be separated off to go it alone, eventually to become Tickford Engineering.

Figure 5.1 Early days at Aston Martin Tickford – the 1982 endurance race engine on the test bed with Trevor Morley. The engine was still on carburettors at this time, it was only used for early shakedown testing of the Nimrod race car in this configuration. The Lucas Mk 2 mechanical petrol injection was used thereafter.

Figure 5.2 February 1982 – Nimrod shakedown testing at Silverstone with Ray Mallock in the driver's seat. This was with the carburettor engine shown in Figure 5.1.

During my time with AMT, I worked on a number of varied and interesting engine projects, including the 1.3-litre Frazer Metro, the 2.8-litre Tickford Capri and marine engines, as well as race and road AML projects. The 1,000 horsepower racing marine engine and the twin-turbo Bulldog are mentioned later in this book.

The AMT Group C endurance race engine project was a three-year project started in 1981 for the 1982 season, and was aimed at the new fuel limit regulations that were to be introduced for the 1984 race. I don't think anyone expected too much from the first year, the car was new and the engine would also be new as a race engine. Getting the two to work together while developing cooling systems, engine mounts and intake and exhaust systems matched to the aerodynamics of the car, etc. – there were bound to be teething problems and there were bound to be engine casualties while developing the car. Nobody goes into racing at this level expecting a free ride. In 1982, the first year of racing, we had our problems – such as the distributor rotor arm breaking due to vibration; this was easily solved and further to that the distributor was moved to the centre of the V for 83. There were also losses during cooling system testing in the car, and one car stopped during its first race at Silverstone due to the timing chain tensioner coming loose. But we had one problem that we didn't expect and that was that we could experience valve spring breakage after six or seven hours of race conditions. It only happened in the car, we couldn't break springs on the test bed even after running the equivalent of two 24-hour races. The only way that we could replicate the failure was to run a cylinder head test rig at 7,800 rpm continuous. At the time the engine was limited to 7,500 rpm. The test rig was based on a cylinder block, with an electric motor driven dummy crankshaft driving the standard half speed chain drive to the camshafts of the test cylinder head. Understandably things were getting a bit frantic as the 1982 race season was already underway, so we ran the test rig night and day to arrive at a solution. Everything was thrown at resolving this totally unforeseen problem, including burning a lot of midnight oil. The eventual solution involved new special surface treatment Schmitthelm valve springs, stronger valves with a reduced dish in the head and stronger top spring retainers. The last two parts were to ensure that the fitted spring length did not open up and allow the springs to go into surge. Despite this concern the car – entered by Lord Downe and run by Richard Williams – did very well in its first season, coming third in the championship. It ran very well at Le Mans that year, running consistently in fourth place for some considerable time until, in the early hours of Sunday morning, the engine started to use a lot of oil and eventually lost a cylinder. But with some careful coaxing by the RSW team it got home in a creditable seventh place overall, the fourth Group C car.

I did an inspection report on that engine when we got it back to AML after the race. I have to say that it is the only time that I have ever seen a completely worn-out Aston Martin engine. Valve guides were worn, piston rings were rattling around in their grooves and pistons were worn in the bores. The crankshaft was not too bad. It seems that – due to some aerodynamic effect – the air intake for the engine was vacuum cleaning the race circuit, taking in road grit to the engine in the process with the inevitable result.

For 1983 the engine was extensively redesigned and lightened. It also became used as a stressed member for the EMKA race car. Further improvements were made to the valve gear, and peak power rpm was reduced to 7,000 rpm in the interests of durability. The EMKA was a beautiful car, built very close to the regulation weight. It was prepared and run by Michael Cane Racing and had a great deal of potential but needed development – it was its first year after all.

Figure 5.3 The EMKA with the AMT engine installed.

So 1984 was supposed to be our year, but our three-year project to win the Le Mans 24-hour race was thrown out of kilter when the fuel efficiency regulations that we had been aiming at were relaxed at the last minute. More fuel available meant more power could be used, more than we could reliably make from our engine in normally aspirated form. So the engine was turbocharged for the 1984 season. I didn't do the development work on the turbocharged engine as I was about to leave Tickford to take up my new position in charge of engines at AML.

AML Engineering Department, 1984

When it came, the AML/AMT separation was a painful experience for both companies but probably more so for AML, as not only had we lost the benefit of having a significant engineering resource but also we were now worse off than before. The Tickford side of things had evolved out of the AML experimental department and had absorbed all of those resources, which – as it was now to become a stand-alone company – effectively left AML without engineers or an engineering department.

At the time I was still working on engines as a development engineer within the Tickford division of AML. Along with a few others I was offered a position within the new AML engineering department, which meant that I had to make the very difficult decision to stay with AML or with Tickford. I had been with AML since the very early part of 1959, and had a strong loyalty for the cars and the company. On the other hand, I had been in at the beginning of the Tickford venture, which had provided me with a much wider engineering experience; also I had a lot of very good friends there, not to mention a great deal of respect for the directors and engineers that I worked with – as I said earlier, they were all top guys in their respective fields. But eventually a few of us were persuaded to jump to the AML side of the line to form the nucleus of a new but much smaller engineering department, with Michael Bowler as director of engineering.

The new AML engineering department had to hit the road running with – from my point of view – a major engine project in progress; a project that I had completed most of the base engine development as part of AMT, but for which I would now be the project liaison engineer for AML. At the same time there was an oil consumption problem beginning to show itself and some customers were concerned about the amount of oil that their engines were using. Oil consumption is a major problem to sort out and, as I had just started, it was decided that this would be best returned to our piston manufacturer to sort out for us. Apart from building test engines for them, this released me to tackle the next major upgrade to the Vantage engine derivative for the Zagato Coupe with a 300 kph target speed. And then the big one, the creation of the new 32-valve engine for a completely new car, the Virage. It was this last project that again brought home the reality of the costs involved in producing a completely new model to conform to the new, more stringent, European emission regulations, costs that would take an awfully long time to recover at a production rate of one or two cars per week, if ever. The twin supercharged Vantage that followed was an even bigger project and the costs were rising accordingly. So the opportunity offered by the Ford Motor Co. must have looked very attractive to Mr Gauntlet when it came.

Changing Times

There were many changes at AML during the 1990s, the most significant being the takeover by the Ford Motor Company, which happened midway through the supercharged Vantage project. People's reactions were very mixed when this first looked like becoming a reality, but there can be little doubt that we had reached a point in AML history where we could no longer go on as we had been. The increasing costs of design and development to comply with the ever-more stringent safety and environmental standards for each new model were not recoverable by making motor cars at the very low production volumes that we were used to – particularly as they were already very expensive and labour-intensive hand-made motor cars built to the highest possible standard with almost the entire car being manufactured on-site. From raw casting, sheet of aluminium or leather hide to the finished article, Newport Pagnell was a true manufacturing plant. It wasn't that AML was a badly run company; quite the contrary, we had achieved a number of industry firsts, many coachwork awards and accolades from the motoring press – such as the fastest car ever tested – and we were well-respected within the industry as a whole, all on a small volume manufacturer's budget. Looking back, it is amazing what we were able to achieve for so little; we certainly didn't need any lessons in economics. Let's face it, the heritage that the people at AML had created was what attracted a major player like the Ford Motor Company to invest in it. And we must not forget that James Bond drove one of our cars – what better recommendation can you have? It was just that times were changing, the exclusivity of hand-crafted coachwork, etc. didn't hold the same appeal as it once did and even if it did, it would always be for a very limited market and we needed to sell at least one of our range of cars at much higher volumes to recoup the increased costs of making them. So we had to adapt if we were to survive. The motor business was becoming much more international with major units being sourced from specialist suppliers, the actual build process was to become much more of an assembly process, albeit to the same high standard.

Accepting that things had to change, they did not go too smoothly at first. There was a particular time early on during the restructuring when the relationship with the new management was at a low. There was a definite feeling of them and us, not a great start for any management team. An atmosphere developed within the factory that, among other things, resulted in the creation of the infamous underground magazine *The Pits*, an extremely irreverent but funny publication. It served to raise a smile at a pretty unfunny time. Most people saw the amusing side of it and looked forward to the next edition, but one or two took it seriously. To this day the identities of the people involved in its publication are a closely guarded secret. Thankfully things eventually started to look brighter and there was no longer a need for *The Pits*. I should say that, however bad things were internally, everyone understood the need to present a united front to the outside world. So there was no washing dirty linen in public.

It is natural for the new owners to want to install their own managers and systems. But a lot of the longstanding AML procedures for recording test results and project notes were officially abandoned along with most of our other in-house engineering change system controls. These procedures had originated during the John Wyre days, so perhaps should not have been dismissed quite as lightly as they were, particularly as we were still developing and producing the V8 cars that were very dependent on the disciplines of the these old systems. Also, all engineering staff except for my department were put into an open-plan office, which was so unpopular that it became known as the Gaza Strip – nobody wanted to go there. Everyone in that office was limited to one filing cabinet only, which meant that a huge amount of our old records had to be dumped. Gone were the days of the dusty old office with wall-to-wall filing cabinets full of records going back to the early days. The result was that loads of AML history was lost in the process. Kept the place tidy, I suppose.

Inevitably there was a gap when keeping records was very much down to individual managers and proper control of the engineering change process was at risk. This situation was further complicated when department managers lost that title. All positions or job titles were taken away below the very top level of new management. One was either an engineer or a production line worker with no distinction between types of engineer or production worker. So officially there were no engineering section managers or chief engineers.

In reality, things had to go on much as before but were a little more difficult due to the lack of any formal structure and the disciplines associated with such. To further cement this working arrangement, the whole workforce was trained in the benefits of separating out into three teams, as opposed to the one team approach practised by AML up to that time. An example of more being better, I suppose.

Appropriate job titles were not reinstated until Mike Morton took over as general manager and some semblance of normality returned. I believe that Mike was on secondment from TWR, where he returned to after his stint at AML, which was shortly after what was left of the engineering department was disbanded except for a very small product support capability. Effectively all further development of V8 cars was to stop, other than to respond to customer concerns. To Mike's credit, he did sit down with each of us to discuss the situation and to offer help and advice where he could. There was no suggestion of redundancy, but most of the younger engineers got the message and left of their own accord soon after.

Before his departure, Mike had worked very hard with the engineering team in preparing various proposals to revitalise the standard Virage model, which by this time was becoming a little long in the tooth. But all of our proposals had been rejected they were either too costly or, if we cut them back, they didn't provide what was required. Later on, when things began to settle down, we did do a bit of further development on the Vantage but only to keep up with ongoing legislation. And we also developed the V8 Coupe, but that started out as a Saturday job, so a story for later.

Looking back, the days of the coach-built Newport Pagnell V8 engine cars were numbered from very early on; rightly or wrongly, there was a general feeling that anything or anybody from the old AML had little value to the new company. And it seemed to take an awful long time with a few *very* unattractive blind alleys along the way before the new Vanquish project was revealed to the workforce by Bob Dover. It was an immediate morale booster – at last we had a positive direction with a worthy successor to the Newport Pagnell V8 cars. Another new chapter in Aston Martin history had just begun. The sighs of relief could probably be heard in Dearborn, which in the end made it all worthwhile. By the time that you read this, the AML that I knew will have been transformed into a very different company with a strong and secure future in front of it. This is what everybody wanted, so undoubtedly the short-term pain and uncertainty was worth it.

Engineering Directors

T here were a number of changes of engineering director up to and during the Vantage development programme. So starting from the beginning of the new engineering department in 1984, after the split from Tickford: Michael Bowler started up the new department and saw the completion of the 1985 model year EFI cars for Europe and the USA, the Zagato Coupe and convertible projects and the 580X specification Vantage. He then moved on to look after the Zagato project in Italy.

Bill Bannard took over engineering and was in charge during the Virage project and the beginnings of the 3-valve Vantage. The Vantage project started in earnest when Andrew Woolner took over from Bill as engineering director. Rod Mansfield then took over from Andrew for a short time and became the last director of engineering at Newport Pagnell AML. Thereafter Bob Watson took control as engineering manager during the initial Ford takeover period and up to the end of the Vantage project.

Vehicle engineering was effectively disbanded shortly after the Vantage went into production, with only a small product support staff being retained. My engine department was reduced to myself, one contract draughtsman, one contract development engineer and one full-time technician, although I still had responsibility for production engine build, the machine shop and the engine test facility, including the occasional hire of that facility.

For a short time it looked like we might get Tom Walkinshaw as director of engineering. This was at the time when it was being decided whether or not we should put the Vantage into production. We had a number of good engineering meetings under Tom's chairmanship. It was refreshing to have someone in high office with an engineering understanding again. I am not sure of the contractual arrangements, but he may well have been our last director of engineering if things had worked out.

Engineers

S hortly after Andrew Woolner became engineering director, having taken over from Bill Bannard, Rob Robinson moved to production quality control. Rob had been engineering manager at the time. And Scott Ellis, who had been chief engineer on the Virage project, left to set up his own business in race car preparation. Towards the end of 1990, Andrew Woolner recruited Mike Booth as Vantage programme manager. This was shortly after Andrew Marsh had left, who had been Vantage project manager up to that time. I inherited the project manager title in addition to my position in charge of engine design and development. That didn't last too long due to other changes within the company, but I remained in charge of all Newport Pagnell engine projects from 1984 up to 2000, when the last V8 engine was produced. Engine development ceased at Newport Pagnell and I went into retirement nine months before my 65th birthday. Nice timing.

It was against this background that the Vantage eventually went into production. Luckily the Vantage project was almost complete when the new management team took over or we might not have made it at all. Even then the project was halted several times, but Rod Mansfield managed to keep it going somehow. And later on I know that Mike Booth had to work very hard to justify putting the car into production. Equally I had to justify the twin-supercharged engine on a number of occasions. To some of the upper management of the time, the service department's 6.3-litre conversions looked attractive as an alternative but – as I explain later in this book – it was unlikely that it could achieve the necessary power output in a world market certifiable specification. So you can see that the eventual launch of the Vantage in 1992 was against a background of some considerable doubt from within the new company. The fact that it was produced at all was ultimately down to the legendary foresight of Walter Hayes. And even then it was thought of as a side-line to production proper.

The Vantage development cars had all been built in the engineering workshop as was usual practice. So with no further engineering projects planned, the workshop space became available and – because of the small number of cars that it was anticipated would be made – it was decided that this practice would be continued to produce customer cars. At the time it was thought that we would make seven cars. A small special-build team was put together for this purpose under the leadership of Mike Booth.

Engines, however, would be machined and built by the normal production process under my care. But the run in and test process would still have to be on the development test bed as the production beds could not cope with the power. This process produced the first six customer cars. As demand for the Vantage increased, it became necessary to transfer it to the mainstream production line. It also became necessary for the engines to be run in and power tested on the production test beds but with the supercharger bypass locked open to limit the power. Eventually we were able to upgrade the development dyno to an eddy current type for better stability during engine mapping, which meant that the trusty old Froude 'F' type dyno could be transferred to no. 1 production bed and fully supercharged testing of production engines could continue.

Inevitably the Vantage became recognised for the great car that it is. It soon became the mainstay of production at Newport Pagnell and continued to be so until the end of V8 engine production in July 2000. In a way it was a shame that it had to be the main product from the factory, it would have been better to keep it as the totally performance-biased car that it started out to be. But because it was our main breadwinner, it had to be all things to all men, which meant that it took on weight in the form of sound deadening, etc. in an effort to refine it into something it was never intended to be. Some latter cars were hitting the scales at 2,300kg – significantly heavier than the Lagonda. An update or replacement for the Virage was well overdue as the main product. Also I was disappointed that the Vantage would never realise its full potential as a world market car, which is what the engine was designed for as part of the original brief. It was a challenge that I was looking forward to. The doubters would just not go that one step further and allow us to certify to the USA standards, which was very sad for the USA. At its release it was the world's most powerful production car and it was still the most powerful production Aston Martin up to the One-77 released in 2011, and would certainly have been the most powerful production car in the States for some time.

Fairly Secret Engines

During those difficult times, I was very fortunate to have had the support of an engine development team of totally dedicated, very loyal and enthusiastic people, a team that I was very privileged to be part of and who made it all come together. They were colloquially known within engineering as 'Fairly Secret Engines' a name that was humorously given to us by the legislation and certification group. I believe that the inference was that we were inclined to keeping our cards close to our chest. More properly we were called AML Engines or, later, AML Power-train.

The team consisted of:
- Tony Bailey, technician/engine builder.
- Roger Date, technician/dynamometer operator.
- James Grantham, development engineer.
- Bev Jones, development engineer and a very brave and competent test and development driver.
- Steve Rawson, senior development engineer.
- Myself, head of AML Engines.

Other key people directly involved in the Vantage engine project were:
- Robin Jackson, production engineer.
- Geoff Hoden, chief inspector.
- Steve Armitstead, electronic systems manager.

Barry Rowledge who had been in charge of the design office for many years, had left the company early on in the Vantage project, so I had to contract out any engine drawing work. Fortunately I knew just the person for this – Len Knowles, who had previously been in the employ of AML and now ran his own company as Milton Keynes Design. Len and I had worked together before, so he was used to turning my rough sketches into proper engineering drawings. And he knew the AML systems and procedures very well.

The reference to keeping our cards close to our chest in the engine department was not entirely without foundation, although this only became necessary in the last few years of Vantage development. There were a number of instances when I would be in a meeting and find that strange engine power outputs were in circulation. What had happened was that someone had walked into the test bed and picked up bits of information by talking to the guys or by reading a power test sheet without understanding what was being tested. The old adage of '2 + 2 = 5' springs to mind. It would have been better if the person in question had spoken to me first. No doubt the figures had been presented on my behalf and with the best intention but without an understanding of what they meant. After a while, the guys cottoned on to what was going on and were not so forthcoming. But they were not averse to leaving the odd doctored test sheet around as bait. I remember finding one sheet showing 800+ bhp, obtained after a dyno breakdown requiring repair and recalibration.

Figure 9.1 Fairly Secret Engines in 1992. Left to right: Roger Date, Adrian Moseley, Arthur G. Wilson, Tony Bailey, Bev Jones and Steve Rawson.
Photo courtesy of Steve Rawson.

While there was no question of keeping anything from legislation and certification, they could see the funny side of what was going on and coined the 'Secret' title for us. It was not really a matter of not releasing information, just that it was best presented in the proper manner by the engineer in charge.

I must mention the guys that built the engines. They are *(in no particular order)*:
- Ron Russell
- Roy Robarts
- Terry Durston
- Chris Bennett
- Michael Peach.

Engine Development Test Bed

The year 1984 had seen the separation of Aston Martin Tickford (AMT) from Aston Martin Lagonda (AML). And, as has already been said, a new engineering department had to be created very quickly at AML. This task fell to Michael Bowler as director of engineering and it was he who offered me a position within this department with responsibility for AML engines.

There weren't many of us to start with. John Lipman was chief engineer, Barry Rowledge ran the design office, Roger Hodgkins was in charge of electrics, Roy Goldsmith had legislation and certification and I was in charge of engines and acting deputy to John Lipman. Later we would be joined by Keith Longmore on electronics, Scott Ellis and Rob Robinson as development engineers and Colin Pinn on admin. John Lipman left soon after Michael Bowler moved to take over AML future products and, in particular, the Vantage Zagato project in Italy. Bill Bannard became director of engineering. I believe that Bill *(affectionately known as Kaiser Bill)* wanted to foster an element of new blood to introduce new ideas for the next-generation Aston Martin. So Scott became chief engineer and Rob became engineering manager. The Virage project began in 1986 and was released at the NEC motor show of 1989 as a world market car. It was a significantly different vehicle from its immediate predecessor, the 16-valve V8 Coupe.

As I will be referring to various power outputs from Aston Martin engines in this book, I think that I should write something about the test facilities at Newport Pagnell. In the past few years, AML had stopped publicising engine output. This was partly because of exaggerated claims by motor manufacturers in general during the 1950s and early 1960s, which were becoming a bit silly. So AML opted out and stopped quoting engine output. During that time, most magazine road tests estimated 400-plus bhp for our V8 engine, whereas in reality we were making around 310 bhp. But we also made a bucket full of torque, which is what was really impressing them. Later on it became part of the certification process for all new models to demonstrate the power output from an engine by a test witnessed by the Vehicle Certification Authority (VCA), and it is those figures that were published.

All of the power outputs referred to in this book will have been obtained using the AML engine development test facility. Originally there were two engine test cells used for what was then the experimental department; both had a Froude DPX water brake. In those days, the operators were in with the engine; usually two but sometimes three engineers would be required depending on what sort of beast was on test – one to operate the engine and the dynamometer and take readings, one to balance the brake and measure the time required to use a measured quantity of petrol for specific fuel consumption calculations. Later on this was converted to one test cell and an observation room with remote control of the DPX dynamometer and electronic fuel measurement, much more comfortable for the operators. At the end of 1979 we upgraded to a Froude 'F' type hydraulic controlled water brake with full electronic control, including a ten-stage programmer that allowed us to make a reasonable replication of various race circuits, all state-of-the-art for the time. This equipment allowed us to achieve VCA approval for in-house witnessed power testing for certification purposes. The facility and all of its equipment had to be regularly calibrated and certified to a standard required by the VCA to maintain that approval.

In the early 90s the Froude 'F' type dynamometer was replaced by a large single rotor eddy current dynamometer for greater stability during engine mapping of the electronic engine management systems. I mention that it was a single rotor unit because of the high inertia involved by such a large high-torque, high-speed unit that had to be specially prepared for us by Froude Consine. I had wanted to install three smaller units in tandem to reduce the inertia effect, which was the more normal route for high performance engine development, but my budget did not allow for this. The high inertia or flywheel affect did seem to have an effect on power output, as the results were always slightly lower than those achieved on the 'F' type although both dynamometers were accurately calibrated and certified.

Power take-off is from the engine flywheel, so the figures in this book represent flywheel horsepower without any of the vagaries of estimating losses through the drive train. A relevant full exhaust and induction system is used for the various tests, as is the relevant standard correction factor for barometer, humidity and inlet air temperature.

On the other hand, the production engine test facility was used purely for engine run-in and sign-off. All of the equipment was still subject to the same calibration standards as the development facility. But the cells were much smaller, originally designed for the six-cylinder engines. The exhaust systems were of a simple, non-tuned design to facilitate ease of installation. Engines were run in to a schedule and then full load-tested up to 85 per cent of max rpm. The figures obtained during the power test were for sign-off purposes only and do not represent the full power potential of the engine due to the non-tuned exhaust and the difficulty in controlling the air inlet temperature because of the small cell and the fact that the engine was not taken up to full power rpm. I felt that it was necessary for me to clarify this, as it did cause some confusion even within the factory and I have heard these figures referred to as being representative of maximum engine power with some authority.

On the subject of the exhaust system, it is interesting to note that the 16-valve Vantage 580x was particularly handicapped by the simple non-tuned exhaust system on the production test beds, as power was reduced to similar to the standard engine of the time.

The sign-off figures for the production test beds were arrived at after running a full power test of an engine on the development test bed and then repeating the test with the same engine on the production test bed on the same day.

When I started at AML, the production test beds were quite new. They were used to run in the six-cylinder engines of the time and power test up to 5,500 rpm. The engines were initially turned over for 20 minutes by an electric motor prior to being started on petrol to begin the run in schedule. Every fifth engine was then stripped for inspection prior to being rebuilt again.

Later on a gearbox test rig was added to the end of the test bed building. It used a DB4 GT engine as the mule to drive the gearbox on test. Load was applied by two large flywheels from the final drive unit. This was at the time that we were building the David Brown gearbox for the DB4. The GT engine sounded wonderful when doing a gearbox test and no doubt encouraged the test driver in his work. However one day there was a slight dampener put on this enthusiasm when the propshaft broke during a test. The operator was protected from the propshaft by a safety guard, but the sound of both ends flailing around just next to him must have really grabbed his attention. The gearbox test bed was eventually decommissioned when we went over to using the ZF gearbox. and later on the experimental engine test beds were built on the site that it had occupied next to the production test beds.

We also had a rolling road test cell in the position now occupied by the service department offices. Quite often we would have customer's race cars on the rolls for final set-up. But the more usual everyday use was for steady state carburettor calibration work in the car and sign-off for the very early carburettor V8 Vantages that were built up in the experimental department.

I mentioned earlier the close family relationship of AML with customers and dealerships. Quite often we would make available the experimental test bed to set up a race engine for an Aston owner for a nominal sum. From memory, I believe that the sum involved was £500 for three days, the last day of which was for my time to optimise fuelling and ignition timing, etc.

When I took over responsibility for AML engines after the split with AMT I found that the production test bed facility had been allowed to deteriorate into a pretty dreadful state, with only one of the three cells in anything like a usable condition. But at that time responsibility for the production beds lay with the production department. My involvement was only required for technical advice or to specify any new equipment. But eventually I inherited full responsibility for that area also, but with the uncertainty surrounding V8 production at that time it was unlikely that I would be able to raise the necessary funding to improve things for the production facility.

However, a number of events came together that enabled me to make some improvements, based on a very modest budget and a great deal of help from the production engine builders, mostly in their own time. The only usable production test bed was not capable of testing a Vantage engine with the superchargers engaged. Initially the engines had to be tested with the superchargers in bypass mode. As I mentioned before, with the change over to the Ford EEC3 engine management system for the Vantage engine there was a need for greater stability during engine mapping which could only be achieved with the eddy current type dynamometer. This made the ex-development 'F' type dynamometer and its control system available to production, which would resolve the supercharged Vantage test issue. By now the Vantage car was firmly in production and somehow I managed to get budget approval for a soundproof enclosure and ventilation system to complete the 'F' type dynamometer installation in bed 1. Bed 2 was demolished to provide an observation area between bed 1 and what had been bed 3, which would now become bed 2.

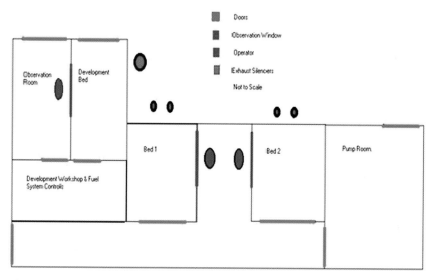

Figure 10.1 Schematic of AML Newport Pagnell engine test facility.

Talking to Tony Batchelor, who was our Ford liaison engineer, it appeared that there could be an opportunity to hire out a test cell to TWR who were responsible for the DB7 project. We just needed to get cell 2 up to scratch. Tony was sure he could help and somehow managed to scrounge a surplus eddy current dynamometer suitable for the normally aspirated V8 engine and the DB7 engine from the Ford Motor Company complete with controls, etc. On this basis I put together a budget proposal for the soundproof enclosure and ventilation system based on the hire fees recovering the costs. Our maintenance department was pretty stretched at the time and could only offer limited support but could supply paint, brushes and pipe fitting, etc. – all I needed was the labour.

The production engine builders came to my rescue; any spare time that they had waiting for parts or whatever they rolled up their sleeves and got stuck in. They didn't have to do it; it was a very dirty job at times, cleaning up years of neglect. The end result was a respectable, if not flash, test cell with a nice little eddy current dynamometer.

1984 Engine Status

A t the time of the AML/AMT separation, AMT was in the process of developing the Weber EFI version 16-valve V8. I was responsible for the engine development, working for David Morgan who was chief engineer. And, as was usual at that time, European and federal versions with a choice of manual or automatic transmission were being developed. This project was to be an industry first for a fully sequential, closed loop engine management system using the new Weber IAW system.

I had completed the base engine work and when I moved back to AML, Chris Bale took over the project at Tickford and I became the AML end of that relationship. The new AML engines group at that time comprised of two people: Tony Bailey as technician and me. Roger Date would join us later to look after the engine test beds. Roger had worked in the experimental department test beds at various times in the past and had at one time worked with George Evans during the early V8 engine development.

During the design stage of the 1985 EFI engine – before I left Tickford – we had made plans for a turbo version. If you look at the inlet manifolds on that engine, it can be seen that they are deliberately set low and are pushed back with forward-facing throttle bodies. Very much with the low bonnet line of the Lagonda and a front mounted twin turbo arrangement in mind, the inlet was designed to function as a normally aspirated or a turbocharged installation. A turbocharged mock-up was made as part of the design process. This was a continuation of the experience gained from the experimental twin turbo Lagonda of 1979. Unfortunately the events of 1984 overtook it and it came to naught.

Figure 11.1 Lagonda engine bay from the front.

Vantage 580X

Having completed the 1985 EFI V8 project with AMT, the next job for the new AML engines group *(both of us)* would be our first full in-house engine project. It would also be the final development of the 16-valve V8 engine. The result was the 1985 580X specification Vantage. This was partly based on an earlier special project that we had done for a Mr Ubel in South Africa, who had a problem with impaired performance from his Vantage due to the altitude. This was a problem that was exacerbated by the turbocharged Porsches of the day, which were less affected. But there was also a very strong influence from the Aston Martin Tickford endurance race engine of 1982/3. The engine for Mr Ubel's car had been referred to as the South African specification, but when we released the 580X it was so-called to avoid any reference to South Africa, which was generally out of political favour at the time. Also we wanted to indicate a further step in development from the original South African specification. We could perhaps have been a little bit more imaginative with the type name!

The objective for this project was to achieve 300 kph with the new Zagato-bodied Coupe and to power a new version of the existing Vantage car. The later convertible version of the Zagato would use the standard EFI V8, which allowed for a much prettier bonnet line. The 580X engine specification had been initially released to our service department as a conversion option for customer cars. This was while we worked on an emission calibration that would allow us to put it into production, not an easy job with such a highly tuned engine. As such, it was the last non-catalyst engine to be developed and was good for 437 bhp and 400 lb/ft of torque. Gaining emission certification for the 580X had been hard work and we knew then that more stringent exhaust emission legislation was just around the corner, which would inevitably mean the end for our V8 engine as we knew it. But it had served us well since 1969 so if it had to go, the 580X would be a good note to go out on. Automatic and manual transmission variants were developed.

Figure 12.1 Zagato engine bay showing air intakes that mate up with NACA ducts in the bonnet.

For a short time, special carburettors were offered as an option for the 580X engine. These were made from the standard Weber 48IDFs by boring them out to 50 mm. This operation was always a risky business due to the possibility of breaking through the carburettor casting. Later carburettor castings were even thinner, which increased the risk and eventually made it uneconomic to continue with this option. They were always a handmade option made by Weber and myself and only created to special order by AML. It was a very labour-intensive process. A particularly tricky part of the operation was that the old progression hole position had to be completely removed and plugged before the new progression holes (four instead of five) could be hand-drilled into the correct positions.

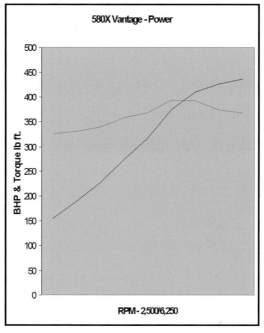

Figure 12.2 580X Power Sheet 1.

At the time of the release, the 580X engine was promoted as having been developed for the limited edition Zagato Coupe. The fact that the same engine went into the Vantage was less well publicised, which is a shame as – to my knowledge – this did not prompt the need for a full magazine road test of a 580X Vantage. The figures would have made very interesting reading, bearing in mind the hefty power increase over the already very quick earlier Vantage – the one mentioned earlier in this book that *Motor* magazine considered the fastest car they had tested up to that time. Most magazines still quote the performance figures for the earlier car.

Figure 12.3 The Zagato prototype at Le Mans with Roy Salvadori at the wheel. Roy Salvadori was there to do a demonstration run with the car. In the event, the engine did not run well due to fuel starvation caused by a particle of bodywork debris that had become trapped in the fuel tank breather. Remember, this was the prototype car on which all the bodywork styling had been done. I was very embarrassed and annoyed at this, as one way or another I had been prevented from doing my usual pre-event performance test and check of the car. This was despite Bill Bannard's insistence that the car and suitable backup be made available to me.

When we got back it only took me one squirt down the Bruntingthorp test track to confirm the problem and no time at all to cure it. From Roy Salvadori's description at the time, I had gained a good idea of what the problem was at Le Mans. Among other things (just to be sure), I had provided an alternative tank vent for a second run. But unfortunately there was no time for a second run.

As I have said, prior to our achieving full European homologation for this engine specification, a number of customer cars were modified to 580X condition by our service department. These cars were referred to as having the 'X package conversion' or being an 'X pack Vantage'. Production 580X cars should not be referred to as such, as there was no conversion package involved. The term 'X pack' being applied to the service department conversion slightly bothered me as I can't help but remember it as having previously been used to identify an upgrade treatment for a Ford Capri. Not that there is anything wrong with a Capri – I had one once and it served me very well.

There was one more development of the 5.3-litre 580X engine in cooperation with the special build section of works service. This was done as a one-off for a special Zagato Coupe shortly before I retired from AML. Designated 580XR, it produced 482 bhp at 7,000 rpm and 412 lb/ft at 5,500 rpm running on unleaded petrol. Compression ratio was standard at 10.2/1. I had wanted to increase this to 11/1 but the pistons were not available in time; 11/1 compression could have given us the magic 500 bhp.

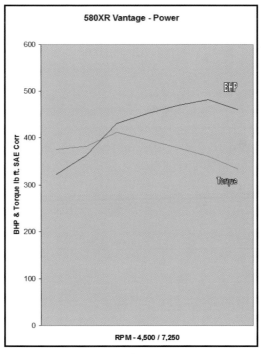

Figure 12.4 Engine built by Ray Brown – AML works service.

Four Valves Per Cylinder

By 1985 the 16-valve engine in its federal configuration was reduced to producing around 260 bhp and 321 lb/ft torque even with the new Weber injection and engine management system. This was due mainly to the low compression ratio required to control the production of oxides of nitrogen (Nox) during the combustion process and the exhaust manifold design restrictions imposed by the close-coupled catalysts that were required to clean up the exhaust emissions. Environmental concerns within Europe would very soon dictate that similar legislation would be imposed here. And it was most unlikely that European customers would be satisfied with the performance available from what would be very similar to our 260 bhp federal specification engine.

Nitrogen monoxide (NO) is converted to nitrogen dioxide (NO_2) in the presence of Oxygen. NO_2 is a highly poisonous gas. NO and NO_2 are generally referred to jointly as Nox. Increased peak combustion temperatures can result in increased Nox emissions.

With the convergence of exhaust emission legislation and the imminent introduction of low-octane unleaded petrol, it was becoming obvious that any future engine project would require a more modern engine design if we wanted to be able to produce Aston Martin power and still comply with world market legislative requirements. Things didn't look good. On the plus side, the similarity in legislative standards would allow us to concentrate on the development of one engine specification for all markets – in fact a world-market car. Production costs would also be reduced by only having to manufacture to the one specification.

Other high-performance engine manufacturers were also having to prepare to meet the challenge of the new European emission standards and of the lower-octane unleaded fuel. Unleaded fuel was required to avoid poisoning the catalysts, which were a necessary part of the emission control equipment. There was a lot of work going on in cylinder head design within the industry. The classic hemispherical combustion chamber combined with a domed piston crown that had been so beloved of engine tuners for many years had the disadvantage of an extended shallow crescent-shaped combustion charge with the potential for a long flame front and residual pollutants at the extremities. Within the industry there was an almost universal trend back to a pent roof combustion chamber combined with a flat top or slightly bowled piston crown with three, four or five valves per chamber. The objective was to provide a compact combustion volume with a low surface to volume ratio, and by good port design ensure cylinder charge activity sufficient to ensure good mixture preparation and a controlled burn rate. This was to enable a higher compression ratio to be used with the low octane fuel without creating a Nox problem. Higher compression ratios are essential to thermal efficiency and therefore improved fuel consumption.

As a manufacturer of high-performance cars, AML would need to be at the forefront of this new generation of engine design. So, after some research, it was decided that a four-valve configuration would be the route that we would pursue.

You could say that this was a return in Aston Martin terms, 4 vpc having been used in the 1922 GP cars commissioned by Count Zborowski. But then nothing is new.

Inevitably the cost of such a major engine project would be cripplingly high and a major undertaking for such a low-volume manufacturer. At our production levels of one or two cars a week it could be years before we recovered the outlay for such a project. A number of leading companies involved in modern cylinder head design were asked to quote for the job of designing the new head: an exercise that only served to emphasise the magnitude of what we were about to embark on. We would keep the overall engine design and the development in-house, but even then the total cost would still be pretty staggering. And after we had designed and developed the base engine, there would still be the cost of the other associated engine systems fuel, exhaust, cooling and intake systems, not to mention the engine management system design and the calibration work and final certification in the car.

Callaway Engineering

During the start-up of the new engineering department, Peter Livanos and Victor Gauntlet had met those of us that had stayed with AML and, after an informal presentation of the new plans for the company, we were thanked for our loyalty. This was a gesture that was typical of both gentlemen and the way that things were done in those days. At that meeting Peter Livanos spoke to me about a neighbour of his who had a small machine shop and who had done great things with VW and BMW engines. He had also designed a Formula One engine to challenge the invincible DFV of the time – referring, of course, to Reeves Callaway. Later it would be Reeves who would come up with a novel solution to achieve an affordable four-valve cylinder head design for AML to develop.

The Callaway plan was to use an existing valve and spring pack from another motor manufacturer and to design around it to achieve a compact four-valve pent roof combustion chamber. To protect it for future development, a very rigid cylinder head casting with good flow and combustion characteristics would be designed. Hans Herman had worked with Reeves before on a number of successful projects and it was Reeves' intention to try to get Hans to design the combustion chamber and ports and to provide suitable cam profiles. From the AML point of view, it was thought that hydraulic tappet and timing chain adjustment would be necessary features for a modern engine design. The Callaway proposal was a novel and innovative approach to containing design costs and was well researched in terms of AML background. All in all, it was tailored to accommodate our particular needs at that time.

So it was decided Callaway Engineering would be responsible for cylinder head design inclusive of cams, chain tensioning mechanism, cam covers and fixings. AML engines would be responsible for overall engine design and development. Engine management would be by the Weber IAW system similar to that used on the 16-valve engine. AML would supply a suitable bottom end to enable Callaway to sign off the design and to proof-run the first batch of cylinder heads prior to despatch to us for development running back at Newport Pagnell.

While it was not clear at that time, Reeves would go on to develop a highly competitive race version of the engine *(for AMR1)* while we were engaged in the road version. We were therefore able to benefit from a good two-way traffic in information between both projects and also to feed in some of our previous AML V8 endurance race engine experience *(we would have given our eye-teeth for a four valve head in 1982)*. A good relationship developed between Callaway and AML, communications were very open and free-flowing from both parties making for very amicable project management.

The original intention was that the head should fit onto the existing cylinder block and use the two-valve induction and exhaust system hardware: in effect, a four-valve conversion with the option to retrofit to earlier engines. As our plans evolved, it soon became obvious to me that this approach would seriously compromise the end result. While Callaway were prepared to design within those constraints, it was recognised that this would severely limit the engine's potential for further development. So after I had had long discussions with Bill Bannard *(who had a very well developed bulls**t sensor system, so one had to be very well prepared for such discussions)*, a decision was made to consider the engine from a wider perspective. One of the first things that it became necessary to change on the cylinder block was the cylinder head stud pattern. The stud spacing on the 16-valve engine was closer on the end cylinders than on the inner pairs, 1.94" from cylinder centre line against 2.407". This had caused problems with gasket clamp loads at very high output in the past, so there was already a good reason to change it. For the 32-valve engine, it would limit the space available for the bifurcated inlet and exhaust ports between stud bosses. To alleviate this, the front and rear stud spacing would need to be opened out from the original 1.94". A study of the existing pattern work for the cylinder block casting revealed that with some modification to the internal core the end stud spacing could be increased to 2.2" from cylinder centre line, which would provide more room for the ports in the cylinder head design. Initially this modification was done by hand-rubbing the sand core to produce the prototype cylinder blocks.

The downside of this design change was that we would lose the two rear oil drain channels and front oil feed to the camshafts in the process. This was the beginning of a series of events that would mean that block changes would be significant to the point of requiring a new block casting. It was soon to become evident that what we were getting into was the design of a new 32-valve engine – albeit based on the crank and the Lagonda oil pump of the well-proven 16-valve – not a replacement cylinder head as had been originally intended. A new oil feed was created directly from the centre of the main gallery up to the head face between the centre cylinders on each bank. Regarding the rear oil drain channels, it was thought probable that the other oil drain routes would be adequate. But this would be investigated during engine development and external pipes used as a way out if needed, at least for the first engines. The long-term solution would then be to cast it into the block but this would involve significant changes to the rear of the cylinder block.

It was important to minimise the changes to the cylinder block. As the cost of new pattern work for a new block casting, was not something that had been allowed for, some clever design work was required to adapt the existing tooling to produce the 32-valve block but still be able to produce a casting that – with various bits machined off or not machined – could recreate a service replacement 16-valve block. This is why the 32-valve block still visually resembles the 16-valve and why we have what is an add-on oil drain for the supercharged Vantage cylinder head. The design for the new block would be done at AML in conjunction with the cylinder block casting supplier.

As far as the cylinder head was concerned, the round inlet ports to accommodate the round 2 vpc manifolds went ahead, although the manifold was not used beyond the initial stages of development. It didn't work very well and it made the engine much higher than it needed to be. Later we would discover that the round port was not a good feature of the cylinder head design. To allow for some flexibility during development, provision was made for head-mounted fuel injectors a water gallery outlet under each inlet port and external rear oil drain back to the crankcase. Evidence of the various mounting faces can be found on early Virage head castings. Provision was also made for four-stud exhaust manifold fixings in anticipation of a turbo version. None of these features were used in production, although the injector position was used initially during development. The internal cylinder head oil drainage was found to be adequate for the Virage.

Figure 13.1 First 32-valve engine on the Callaway test bed. The blanking plugs can be seen in the injector positions in the cylinderhead. Note how awkward the induction manifolds from the 16-valve engine look on this engine.

Callaway were very aware of the implications of pattern work costs to a small-volume manufacturer, and were innovative in the use of such equipment. Typically a non-handed core was employed for the main cylinder head casting with end cores to set the timing case at one end or the other, dependent on which hand was being cast. This meant that only one set of main pattern equipment was needed to produce both heads. Evidence of this can be seen in the form of the water jacket blanking plate cum heater take off at the rear of each head. This being the unused opposite end water gallery exit that would have otherwise connected to the thermostat housing at the front of the block. At one time an entirely non-handed head was considered, with add-on timing cover and rear cover, but this was not pursued due to concerns about oil and water leaks via the numerous joints involved. Alan Brett at Birmal – the company that had been our aluminium casting supplier for many years – was a great help during the casting design stage and was very keen to try new casting techniques to cast in the oil feed and air injection galleries. Grainger and Worrall, who now supply our V8 engine castings, designed the pattern equipment for the new head. A major step forward was that we could have a cast in air injection rail including the porting to direct air onto the back of the exhaust valves. We would also have a fully cast finish combustion chamber, which was a new experience for us. It would mean that we were dependent on the accuracy of the casting foundry to achieve balanced combustion chamber volumes. We did have some problems with that one but we got it right in the end. I believe that the cast in air rail was a first for anyone at that time.

Air injection into the exhaust port is used for the first 100 seconds during cold engine start-up. This is to speed up the catalyst light off. Catalysts do not work efficiently at mid-bed temperatures below 350 degrees Celsius. so it is important to achieve this temperature as soon as possible after the engine has started to control the exhaust gas pollutants. By adding oxygen to the exhaust gas, the exhaust gas continues to burn (oxidise) down the manifold, resulting in reduced carbon monoxide (CO) and hydrocarbons (HC) and a higher temperature gas entering the catalyst. Gas temperature is also maintained by running slightly retarded ignition timing during this period. This strategy is used by AML to enable us to retain a performance-tuned exhaust manifold design and to avoid the alternative close-coupled catalyst design, which can seriously compromise the manifold configuration and, most importantly, engine performance.

As had been agreed, Callaway would supply a design and prototype parts for the cylinder head with cam covers, front covers and a method of timing chain adjustment, as well as a piston crown profile to complete the combustion chamber geometry. The rest of the piston would be designed jointly between AML and the supplier. As the design stage progressed, Callaway would also do some preliminary running tests and supply fully assembled and run prototype heads for development work at AML. The first engine to run was at the Callaway headquarters in Connecticut USA on 17 March 1987, the bottom end having been built at Newport Pagnell. I went over for the start and was very impressed by the Callaway layout and the enthusiasm of the workforce. There were teething problems with the timing chain tension that upset the cam phase signal to the engine management system and delayed the start. And the compression ratio was much too high, requiring high-octane fuel for the run. But otherwise everything worked well and showed promise for development back at AML.

Back at base, Renold Chains were contacted to discuss the chain tension problem. Reeves had already contacted them from the USA. They had a very compact hydraulic tensioner and recommended its use with a slipper type chain guide mechanism. This meant a complete redesign of the timing chain system with new chain lengths and guide assembles, as well as an oil supply and a pivoting slipper for adjustment. Design and casting changes were needed to the front of the cylinder block and to the new heads. But existing castings were welded up and re-machined for development. By this time we were well into

Figure 13.2 The original design for the chain tensioner layout – the cam phase signal for the engine management system is driven from the front of the inlet cam.

Figure 13.3 New design chain tensioner system. Note this photo is of a Vantage engine as can be seen by the different cam phase sensor plate on the front of the inlet cam. But other than that the principle was the same for the Virage.

the design stages of what would be the new engine, so I was more than pleased when the engine section expanded to three when we were joined by a very clever young engineer named Andrew Marsh. One of Andrew's first jobs was to work with me on a drawing layout to fit the new chain run into the timing case and make the adjusters fit under the intake system, particularly the left-hand one. There wasn't a lot of room and we spent many hours struggling to get it all in, although it looks easy now! Later on, after we had tested the design during the initial development runs it was given to Ernie Cheshire in the design office to finalise and detail.

The chain tensioner shoe (the curved bit in the picture above with the Nylatron lining) was originally cast by the lost wax process in aluminium, but we found that they broke in half during our early testing, so we had them made in cast iron by the same process but they still broke. The very early development parts had been fabricated in mild steel and had been OK. We did some tests with Perspex front covers to allow us to observe the chain movement during various engine conditions and the shoes seemed OK after our first day of testing, but they broke soon after a cold start the next morning. What was happening was that the adjuster teeth were too fine and were adjusting up the tension as the engine cooled down and the chains became slack. As the engine warmed up and expanded, the chains became over-tight, causing excessive loads on the shoe resulting in the breakage. Reynolds supplied us with an adjuster with more clearance in the ratchet mechanism and the problem was cured. But we stayed with the cast iron shoes.

32-Valve Engine Development

Having received our first pair of heads from Callaway, we started the first engine at AML in May 1987. After a few shakedown runs to establish confidence, a number of power loops were run to establish a safe spark plug heat range and power map for the engine management system. The safe map for fuelling and ignition advance would be the base engine setting for the Weber engineering console, which would then be used to optimise spark advance, fuel pulse width and injection phasing for each test point during development.

Our first runs were with the old 16-valve induction system; a repeat of the Connecticut run. But the intake manifolds looked very big and awkward on the 4vpc heads, partly because the new heads were so much smaller but also because of the difference in port angle, which was 15 degrees from head face for the 16-valve and 34 degrees for the 32-valve – they needed to be more in keeping with the rest of the engine. By now the design for the new car was beginning to evolve and it was obvious that the inlet manifolds would need to be very compact to fit under the low bonnet line, so I had to come up with a new compact intake system to fit within the 'V' of the engine configuration. This would have tuned length crossover manifold trumpets enclosed in a single plenum with the twin throttle bodies at the front. Intake trumpet lengths were optimised on the test bed to provide a wide spread of torque combined with top-end power and the internal volume of the single plenum design was slightly increased over the twin plenum of the old type, which we were aware might cause engine response problems at idle and low speed. We made efforts to reduce the volume by partitioning the single plenum chamber to create a twin plenum, but this resulted in poor distribution and a loss in power output. We had been through similar problems during the development of the 16-valve engine induction, so decided to stay with the single plenum and sort out any idle problems when the time came.

The injectors at this time were still in the cylinder head position, but there were indications of wall wetting in the inlet tract due to the spray angle and the close proximity of the port bifurcation. Manifold-mounted injectors were tested for power and gave a small but useful improvement, particularly in the lower rev range and max torque areas. It was reasoned that emissions could only be better due to the better atomisation of the fuel spray being directed onto the back of the inlet valves. We were able to confirm this during the in-car testing at a later date. The practicalities of mounting the injectors into the manifold for production were a bit more of a problem. Bearing in mind that, by necessity, the manifolds were inside of the plenum and that the injectors and fuel supply would be on the outside. The plenum needed to be airtight other than via the throttle plates. The resulting design involves a number of O ring seals and some very accurately machined components. A neat and, importantly, very compact solution was achieved, if not entirely to the production tolerant teachings of Taguchi.

Air Filters

Connecting the air filter trunking directly to the throttle bodies of the new induction system tended to extend the tuned length and upset the good work done on the intake trumpets. As there were not many options regarding the position of the filters in the car, we needed to break the influence of the air filter trunking at the throttle bodies or as near as possible. Another consideration was that we also needed to provide a means of sound damping to the induction if we were to pass the drive-by noise test. The sound insulation qualities of standard intake trunking are not very good, so a secondary plenum was designed incorporating an acoustic foam covering and an internal shape that would complement the primary intake system, which was what now connected the air filters together and to the engine and acted as a sound-damping chamber. The connections from the secondary plenum to the engine plenum needed to be as short as possible for best performance.

Exhaust

Optimum primary lengths for the four into one exhaust manifold system worked out at 40" but it became necessary to shorten these to 30" to accommodate the catalyst light off and to maintain low-speed running temperature in the catalysts. Historically the AML V8 has used a cruciform section or crossover aft of the manifolds to fill in the mid-range and top-end power. It also gives the exhaust a more even beat, as opposed to the usual V8 sound. The same formula was applied to the 32-valve V8 with similar results: a cruciform sectional area equal to 0.725 of the combined sectional area of both entry pipes being the optimum. Having established the optimum configuration for power, a silencer box type cruciform was developed using the optimised pipe configuration as the power baseline. The outcome is a deceptively simple-looking box, with the entry pipes deliberately terminated closely inside the front end plate feeding into a common sound damped chamber. The exit pipes are treated in a similar fashion to the entry pipes by *not* being extended into the expansion chamber. It is not quite as good as the baseline run with the un-silenced but performance – optimised cruciform pipe, but it's very close. I mention this, as the cruciform box is a very sensitive part of the exhaust system; downstream of this is less sensitive other than to system backpressure.

Camshafts and Compression Ratio

The cams that we were running were 0.390" lift to a Callaway profile. Contrary to the 16-valve engine, the inlet and exhaust cams were the same. The 16-valve engine always favoured more lift/duration on the inlet side. Full open timing was optimised at 110 degrees after top dead centre (TDC) for the inlet and 112 degrees before TDC for the exhaust.

Initially we were running on 91 Research Octane Number (RON) fuels as it was believed that this would be necessary to conform to world market requirements *(Europe and the UK were still dithering between 91 and 95 RON)*. On this fuel, the engine was seriously detonation limited. I think that the USA had a different approach to applying RON and Motor Octane Number (MON) in terms of fuel quality. Callaway were certainly surprised at how poor our fuel was, which may explain the over-optimistic compression ratio of the original cylinder head design. I believe the term 'salad dressing' was used to describe our petrol.

Although 95 RON made the situation a little better, detonation was still a problem. As head of legislation and certification, Roy Goldsmith was asked to conduct a world survey of fuel available for our intended markets. And on the basis of his report it was decided that we could develop the engine to run on 95 unleaded gasoline, which helped the situation a lot *(at the time Australia looked like the only country likely to retain 91-octane fuel)*.

From the first engine run in Connecticut, the compression ratio had been reduced in three stages from the original 11/1 to 10/1 and finally 9.5/1, at which point we became concerned at not meeting the power objectives for the project. So far the compression ratio had been reduced by making the bowl in the piston deeper, but there were design limits on how deep we could go. We really needed more of the combustion volume to be in the cylinder head. The problem was that as the bowl got deeper, the piston crown had to be thicker, which increased the weight and thermal mass at the top of the piston. I had decided that a larger diameter shallow bowl would be the best solution, although this tended to slightly reduce the effect of the squish areas designed into the combustion chamber as protrusions between each pair of valves. Callaway was not happy about this, but understood our problem. Although it was a totally different problem, Callaway was experiencing high speed detonation problems on the race engine and had experimented with removing the protrusions. This had stopped the detonation but at some cost in power. We also tested with the protrusions removed but it made no change to our mid-range detonation problem or discernible difference in power. For interest, the protrusions are equal to 1 cc of chamber volume.

With the compression ratio at 9.5/1 we still had the detonation problem at max torque *(peak BMEP)* and max power was only just on target: we needed to move the max torque point up range to reduce the critical BMEP. To do this I decided to run a longer-duration cam profile that would reduce the effective compression ratio in the mid-range and improve the breathing at higher engine speed, thus moving the max torque point further up the speed range and easing the detonation situation. It would also recover some of the top end power that we had lost by reducing the static compression ratio. Camshafts with a 0.4" lift with a 250 degree period were used with the same 110/112 degree full open timing as the 0.39" lift cams. This regained the power and got the detonation under control. If the power was held for long periods in the sensitive areas to heat things up it would return.

So we were still on the edge but a small back-off on ignition advance would control it. Nevertheless the indications were that, as a road car engine with clean cam timing, we were already getting close to the limit for this head design.

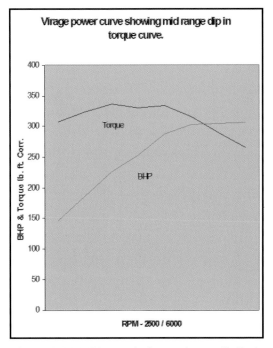

Figure 14.1 This graph shows the small dip in the torque curve caused by the retarded ignition timing required to avoid detonation. Despite this small flaw, the torque curve was still pretty impressive. Particularly at low speed with more than 240 lb/ft available at 700 rpm!

Another issue in this area was that idle quality was becoming a concern and it would not normally be expected to improve with more valve overlap, so cam-timing optimisation was revisited with the 0.39" lift cams in the exhaust side. But the best performance combination was with the 0.4" lift cams in each side, timed as before.

In-car development of the idle quality showed that the main idle problem was an idle surge of about 100 rpm either side of base idle speed. Running more valve overlap was not helping, but we needed to do that to avoid the high BMEP detonation point. So our attention returned to the induction plenum volume. The engine management system was the Weber IAW system, which is an engine speed and inlet manifold air density-based system. The engine speed is picked up via a sensor on the front of the crankshaft and the air density via a sensor in the plenum. The signal response from the air density sensor to the engine management system would be influenced by the effect of the plenum volume. To prove a point, tests were run with the smallest possible plenum chambers attached to each bank of four intake trumpets. Surprisingly this had little or no effect on idle stability. No doubt some work on the calibration would have achieved something, but indications were not good and in any case that size of plenum would have had a disastrous effect on power. An inlet port throttle arrangement with eight throttles – one for each cylinder – was quickly designed and manufactured in the form of a log down the centre of the manifold crossover. Tests with this on the test bed engine gave a rock steady idle with very sharp throttle response, more like a carburettor car. Unfortunately no power readings were taken at that time due to complications with the throttle linkage, which we did not have time to resolve. In the time taken to design and manufacture the eight-throttle system, Weber had made progress with the engine management system using the single plenum system and the idle surge was now under control. So the eight-throttle system was put on ice as something for the future.

In-House Manufacture

By this time we had started to machine our own cylinder heads and a number of manufacturing problems were to emerge. The first heads machined by Callaway had suffered from seat distortion so seats had to be re-cut after running on the test engine prior to shipment to us. We had similar problems, so investigated the interference fit that we considered to be high. This was reduced from 0.0045" to 0.0035", which cured the distortion problem. Eventually we had to change the valve seat material to a more modern sintered material due to inclusions and hard spots that showed up during the seat machining. Also we were still having to hand-lap the valves due to concentricity problems with the pre-finished valve guides. The eventual answer was to have guides made to our own design with an allowance for machining the valve stem bore at the same time as cutting the seats.

A more serious problem was with the valve seat position relative to the combustion chamber surface. Due to the contoured nature of the chamber surface, any minor variation in chamber positioning could result in the seats being proud or recessed at various points around their diameter. This created the potential for HC traps or hot spots, etc. The casting and machining processes were both looked into and improvements made where possible. But we eventually decided that we were putting unrealistic demands on what was possible with a cast-finished combustion chamber. The best heads had been selected for the emission calibration development car at the Weber factory in Italy. In anticipation that we could resolve the problem before we went into production. Eventually it became necessary to redesign the combustion chamber to provide a more regular surface for the seats to be set into. The rest of the chamber was designed to blend smoothly into the outline gas face features of the original design. The intention was to minimise the impact of such a drastic change at what was a fairly advanced stage of the calibration work. I said nothing to the people at Weber, who were well along the way with the emission calibration, but they spotted the change immediately. Nox emissions had dropped noticeably. So, although it was not a totally unwelcome change, it did upset the balance of the calibration in terms of CO and HC versus Nox, which is always a bit of a balancing act. So it meant extra work for our friends in Italy. We didn't have time to quantify it but the feeling was that the detonation situation was also slightly better with the new combustion chamber.

A lean drive cycle calibration can reduce CO and HC, but can also increase peak combustion temperatures resulting in more Nox, a balance of CO, and HC, against Nox has to be arrived at. A point to remember when someone tells you they can reduce your CO.

At the time that I was redesigning the combustion chamber Peter Rang our computer-aided design (CAD) office manager was involved with a company offering full 3D modelling capability. It was decided to use the combustion chamber as a demonstration subject for this system. Peter was able to interpret my changes onto the CAD system and a hard model was produced. There were a couple of iterations of this for fine-tuning and then the data was transferred to the casting foundry to produce a new pattern for the combustion chamber. This was the beginning of all AML drawings being transferred to CAD; up to that time they had been proper pencil and paper jobs.

Engine Finish

Reeves had created a very neat package for the cylinder heads and I like to think that the rest of the engine continued on the same theme. The cam covers had an anodised finish and neat stainless steel fixing nuts. We carried this theme over to the plenum chamber and fuel rail cover, etc. It all looked very smart and we were very pleased with it. But as time passed and we got further into our production run there was a snag. Slight batch to batch colour variations in the anodised finish became a problem that made it necessary for parts to be selected for colour match during engine build. We just couldn't find anyone who could provide a consistent finish. So the anodised finish would not be a feature that would be continued on the next engine design, whatever that may be. The Vantage was just a twinkle in the eye at this time.

With full 50-state federal compliance and European noise regulation compliance in a world market car the engine produced 306 bhp at 6,000 rpm and 345 lb/ft at 3,700 rpm. It compared quite well with its 16-valve non-catalyst and four-star fuelled European specification predecessor and was significantly better than the 265 bhp of the old federal specification V8. So all objectives had been achieved and we now had a single engine spec for all markets. This meant that as far as most of Europe was concerned, the Virage was released equipped with catalyst and running on three-star unleaded fuel before it was really necessary, so initially it had to compete with non-catalyst cars in these markets. But in the USA and similar markets, the extra performance would be very noticeable. Vehicle weight had increased by around 200 kg, which inevitably absorbed some of the extra power.

> Engine men always moan about vehicle weight.

1987 Vantage

With the first standard version of the 32-valve engine completed, thoughts turned to a higher performance or Vantage version. Traditionally we have produced a more performance-biased version of the current standard car of the time for customers who are prepared to accept a little less refinement in the interest of higher performance, the higher-performance version being known as the Vantage. The standard car at the time was the Virage, for which we had just developed the first 32-valve engine using the Callaway design cylinder head. The Virage engine was still very new, but we had explored the design limitations pretty thoroughly during development. And, while we knew that there was some potential for further development, the massive increase in power envisaged for the Vantage would require a substantial redesign.

The objectives set for the Vantage were pretty stiff for the time. World market compliance, good fuel consumption, 300 kph/186 mph road speed and a sub 4.0 second 0–62 mph all to be achieved using three-star unleaded petrol.

The 580X Vantage Zagato had a 300 kph capability, so I suppose a next-generation Vantage could not be slower even if it was world market compliant. Nevertheless, it would be a pretty tall order. The predicted 460 bhp that would be required involved a 45 per cent power increase to an engine package that we knew was already close to its limit. And as time went on, the 460 bhp requirement was increased to 500 bhp, so things did not get any easier. What we were contemplating was the most powerful standard production car engine in the world with full world market certification.

Andrew Marsh had been offered the position of project manager for the Vantage and had done some performance predictions that indicated a requirement for 460 bhp to achieve the 300 kph objective. This tied in with the Zagato Coupe, which could just achieve 300 kph.

Andrew had discussed the move from engines with me before accepting his new position and we had both agreed that it was an opportunity that he could not refuse, although I was very sorry to see him go. His replacement was Steve Rawson, another brilliant young engineer who ultimately would play a major part in the Vantage engine project. I had been asked to find a quiet place and lay out a programme that would complete the engine project over the next six months. So Steve's first job was to come to my home and between us we came up with a plan. Common sense will tell you that the objective was pretty unlikely to be achieved, but on paper it was possible. The fact that we were asked to do it shows the initial urgency placed on the project. The fact that it took so long after that event to produce a customer car gives some measure of the effect of the changes occurring within AML during that time. Just imagine the impact the Vantage would have made had we been able to release it a couple of years earlier than we eventually did, which should have been entirely possible.

At about the time the decision was made to change to the Ford EEC4 engine management system, Bev Jones joined us. Bev had some test bed experience, which would mean that I would not have to spend so much time in the test bed. But his main contribution was to be in in-car engine systems development where his driving ability and background in race car preparation was invaluable. He also made a major contribution to the whole vehicle dynamics development. At this level of performance, the combination of a good engineer and test and development driver is essential.

James Grantham joined shortly after as a development engineer. But sadly he soon became one of the first casualties of engineering cutbacks that occurred after Andrew Woolner left. We also lost our engine builder Tony Bailey and our apprentice Adrian Moseley soon after.

Towards the end of the Vantage project, Bev decided that he wanted to leave AML. The changes going on at AML at the time were not helping morale. Bev's input to in-car development was at a crucial stage. Recruiting a suitable replacement at such a late stage of the project would have been very disruptive, particularly to such a small team. In fact any replacement would have been unlikely at that time due to constraints on engineering budgets. So I came to an arrangement for Bev to support the Vantage project on contract for two days a week until the project was completed, at which time Bev left. Things became very uncertain at AML after the Vantage, and Steve Rawson left soon afterwards.

After that the engine group was reduced to serving a product support role, and any rise in workload was covered by engineers on contract to my department. Brian Cox was one such engineer. Brian eventual became a permanent member of our small team and was a key figure in the development of the V8 Coupe engine.

By now AML Powertrain had become known as AML V8 Powertrain. To identify a new direction away from the in-house product to the new V12, this would be designed and developed off-site. In-house engine design and manufacture at Newport Pagnell would cease with the end of the V8 engine.

Vantage Target Objectives

The computer analysis conducted by Andrew Marsh indicated the following engine performance required to achieve the 300 kph max speed with a sub four-second 0–62 mph vehicle performance.

- 460/500 bhp with torque in the region of 480/500 lb/ft over a wide range to achieve the 0–62 target and to provide high-gear cruising for good mpg and driveability. The 0–62 target was more likely to be dependent on vehicle weight and traction than anything else.

- Max power would need to be at 6–6,500 rpm to ensure target objectives for vehicle speed. Operating range would need to extend to 6,500 rpm to accommodate a vehicle range 0–300+ kph within a normal final drive ratio (FDR). The six speed ZF gear box has a very high (0.5) overdrive sixth so it was always intended that the car would be geared for max speed/max power at 6,500 rpm in (0.8) overdrive fifth.

- Compliance with world market regulations.

At this stage the Vantage was envisaged as a two-seater in the style of the DB4 GT or the more recent 580X Vantage Zagato. The full three-piece suite treatment came later.

Accepting that the Vantage was always going to be a heavy car, it would be important that the engine tune be capable of producing torque as well as bhp. And that peak power should be at an engine speed sufficient for high vehicle speed within normal final drive ratios. The torque curve would be important not only in the peak reading but in the area under the curve. Broadly speaking it is this torque curve that accelerates the car and the bhp that achieves the maximum speed. A balance of mid-range torque (acceleration) against out-and-out bhp (speed) has to be made.

Power and Torque.

Torque or turning force of the crankshaft is naturally independent of engine speed whereas bhp is measured as a function of engine torque and crankshaft speed.

$$\text{BHP} = \frac{\text{rpm} \times \text{torque in lb ft.}}{5{,}252}$$

Power is the rate of doing work, so an engine with the capability to produce its peak power at high rpm will produce high bhp. Our V8 engine has an individual cylinder capacity of 667.7 cc with a bore to stroke ratio of 1.176. In a state of tune conforming to modern legislation, it generally produces peak power at 6,000 rpm or slightly over. Engines with larger cylinder capacity can produce higher torque figures but generally lower bhp due to a tendency to produce peak power at lower rpm. Peak power rpm is reduced due to the time required to fill the large cylinders. An excellent example of this was the long stroke (95 mm) 6.3-litre conversion carried out by our service department. With a cylinder capacity of 793.58cc and a bore to stroke ratio of 1.0842, this conversion was based on the 580X vantage engine and provided a very useful increase in mid-range torque. But peak bhp was actually lower than the standard 5.3 litre 580X due to the power peak occurring at a lower crankshaft speed. Most people that have driven the 6.3 conversion cars will find that hard to believe. But then we are only considering the peak power; naturally the power is increased in the mid-range in line with the increased torque. In the real world, peak power is not used that often, its importance being mainly associated with maximum vehicle speed. So if we accept that all of the power produced by the 580X is required to achieve the 300 kph maximum speed of the car then it is reasonable to assume that the 6.3 version would probably have a lower maximum speed in line with the reduced peak power and the reduced engine speed at which it is produced. As with most things in life, you have to decide what you want. In a real-life situation where you rarely have the need to run to 186 mph, the 6.3 conversion is a much quicker car in the mid-range due to its significantly increased torque.

The original 6.3 conversion based on the 580X was a result of a Service department project in cooperation with Richard Williams. AML engineering did a similar project with the service department on the 32-valve Virage engine. Tickford were later involved with the service department in a number of upgrades on this engine. No production versions of the 6.3 litre engine were ever produced.

Engine rpm range has to be considered as part of the overall gearing when considering high vehicle speed: 200 mph/6,500 rpm = 30.8 mph/1,000 rpm. Having an engine that peaks at a lower engine speed would require a higher FDR to achieve the same speed. Higher gearing would reduce the flywheel torque transmitted to the road wheels, which would need to be compensated for by increased output from the engine. Conversely an engine that peaks at higher rpm can use a lower final drive ratio, which will allow more of the available flywheel torque to be transmitted to the road wheels and can also compensate for reduced mid-range torque from the engine.

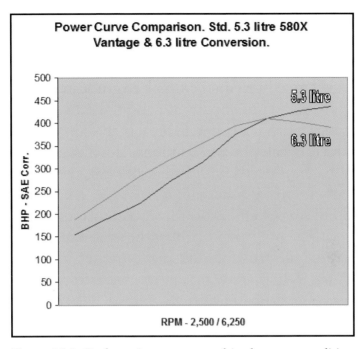

Figure 16.1 Both engines were tested in the same condition, i.e., stabilised readings with the full car exhaust and intake systems, both were fitted with the 48/50 IDF carburettors that were available for a short time as a 580X option. Fuelling and ignition were optimised for the 6.3 version. The 6.3 would have to increase torque by 18.2 per cent at its 5,500 rpm peak to match the peak power of the 5.3 litre engine, or by 12.2 per cent at 6,250 rpm. Whereas the 5.3 would need to increase by an average of almost 18 per cent at all points below 5,500 rpm.

Bhp/Litre/Rpm

When referring to bhp/litre, the number of times that said litre is used has to be a factor in any comparison of engine output. It requires 12.5 kg of air to release the energy from 1 kg of fuel at full load so, broadly speaking, it takes a given weight of air to produce a given bhp. The air relates to the theoretical volume of air displaced by one engine cycle, i.e., the engine swept volume in litres times the number of engine cycles. Thus the capacity of the engine and the number of times that this capacity is used in one minute will define the volume of air processed by the engine and therefore the energy/fuel input and the bhp output. Essentially there is no magic or free lunch. You either run a large capacity engine at relatively low speed or run a smaller engine at higher speed to achieve the same bhp by moving the same air volume.

Obviously this is a generalisation that assumes optimised volumetric efficiency and frictional losses, something that should be considered with the earlier comments in mind. Naturally there are other factors that play a part in choosing an engine specification for a particular application. A larger engine will need packaging space and could be heavier, it will probably use more fuel at idle and light load. A small high-revving engine will struggle to make torque whereas a large capacity engine will make it more easily. Engine breathing and cam-timing systems for high rpm may not be suitable for low emissions and frictional losses will probably increase with rpm. High rpm *(low FDR)* will not improve drive-by noise emissions.

A better understanding of engine capacity might be achieved by multiplying the engine swept volume by the max rpm and dividing by two for a four-stroke cycle.

Taking the 580X max power rpm as max rpm, results in a capacity potential of 16,690.625 litres/min.

5.341 litres × 6,250 rpm / 2 = 16,690.625 litres.

Currently a figure of 12 bhp/litre/1,000 rpm is a good figure for a fully certified normally aspirated production car engine. *(At the time that we were first looking at the Vantage, this figure was closer to 11 bhp.)* There are exceptions — a light car will reduce the need for mid-range torque so greater emphasis can be put on high bhp; conversely a heavy car will need to sacrifice bhp for higher torque — but 12 bhp is a good guide. Obviously this figure can also be used as a guide to the rpm required for a given output from a known engine capacity.

An alternative to higher rpm would be to increase the air density entering the engine via some form of super-charging; again weight of air charge = bhp. This route would not suffer the disadvantages of the state of tune required for high rpm and would package a relatively small engine with associated idle and light load fuel consumption. Torque would be expected to be available over a wide range due to the reduced dependency on induction and exhaust systems tune for cylinder charging.

Note: As I write using the above formula, the 1999 Mercedes Grand Prix engine is reputed to produce 16.67 bhp. The same year, the Maclaren Formula One sports car produces 13.97 bhp. Whereas the early Vantage gave 17.6 bhp and the exhaust-strangled 1997 model year Vantage, with full 75 dB(A) drive-by noise compliance, gave 17.17 bhp.

Full power fuel consumption will be somewhere around ½ pint/bhp/hour by whatever engine route. Using high bhp = high fuel consumption, 1 hour at 550 bhp = 34.38 gallons/200 miles = 5.8 miles/gallon. By which time you would have probably been arrested or have fallen into the sea.

Normal Aspirated

In normally aspirated form, using the 12 bhp standard, our V8 engine would need to run to 8,600 rpm to achieve 550 bhp. Obviously the cam timing intake and exhaust systems required to achieve this engine speed would be most unlikely to comply with world market regulations or result in a torque curve appropriate to a heavy high-performance car. Using the same standard, the 1982 endurance race engine gave a specific output of around 14.4 bhp and ran to 7,500 rpm. But the intake and exhaust were well out of bounds for anything resembling a certifiable road car. Torque peaked at 5,500 rpm, which is not where you need it for a heavy road car. A Formula One engine of the time that we were developing the Vantage (1990) would give around 15.5 bhp.

To run anywhere near 8,600 rpm with clean emissions would really require smaller individual cylinder capacity to reduce the need for excessive valve overlap, etc., which would mean more cylinders to achieve the same capacity. If you can remember the size of the intakes and exhaust systems on the big capacity Can Am cars then you will understand the problems of filling and emptying big cylinders quickly.

Variable cam timing systems were starting to emerge that could be used to clean up and fill in the bottom end. But they were not well-proven, particularly those that could change the duration of the event. Even then the effect was limited and would be a drop in the ocean compared with what we were looking for. Also there is an effect on the effective compression ratio that would mean that the ratio at max power is less than ideal due to having to be optimised to control Nox production at the reduced valve overlap used during the emission test. So what next? Variable compression?

At low engine speeds where it is relatively easy to achieve a full cylinder charge, the longer the valves are closed (less overlap) during the compression stroke, the higher the effective compression ratio, which can provide more torque from the engine. There is also more time for the combustion process to take place, which can result in lower CO and HC emissions. However the higher compression means higher peak combustion temperatures, which can result in increased Nox production. Therefore the static compression ratio has to be reduced to a ratio compatible with acceptable levels of Nox in the reduced overlap condition.

Higher engine speeds require the valves to be open for longer (more overlap) to fill the cylinders, which leaves less of the stroke for compression. To compensate for this a higher static compression ratio is usually used.

> Note: Other methods of controlling combustion temperature can also be used but they are only part of the picture and have to be used in conjunction with the above principles, i.e., exhaust gas recycling into the intake charge is a way of reducing the density and oxygen content of the cylinder charge. This effect can also be used to reduce fuel consumption.

A capacity stretched to say 7-litre would similarly have to peak at 6,550 rpm, which at 875 cc/cylinder would probably not want to do with anything like emission clean cams or induction. Torque would be good but bhp, fuel consumption and engine range would be down on target. Realistically the engine would probably peak in the 420 bhp/5,000 rpm range. So a taller FDR would be needed to achieve the 300+ kph target, which would reduce the power available at the rear wheels and, at the end of the day, it is power at the wheels that counts. So all in all it was most unlikely that our target objectives could be achieved by our V8 engine in an emission compliant and normally aspirated form.

Why Supercharging?

As mentioned before, when it comes down to making power, the air charge is the major component. And the bigger the air charge processed by the engine in a given time, the greater the power output. As has been said, this can be achieved by engine capacity in terms of air movement in a given time (2 litre × 8,000 rpm = 4 litre × 4,000 rpm, etc.). But the larger the rpm operating range, the less likely it is that breathing will be able to be optimised over all of that range. Supercharging of one type or another would process more air for a given engine capacity or operating speeds and would be less dependent on valve overlap allowing for an emission clean engine design in terms of CO and HC. Compression ratio could be reduced to allow for the increased charge density, resulting in lower peak combustion temperatures during the emission test and therefore lower Nox emissions.

To summarise the thinking behind using supercharging to achieve high output and world market compliance:

- Smaller engine capacity for a given output = low fuel consumption off boost = low CO_2 emissions.
- High torque at low engine speed allows higher gear to be used during cruise = low fuel consumption = low CO_2 emissions.
- High torque at lower engine speed allows for higher overall gearing = drive-by noise emissions reduced.
  - High torque throughout the engine range = driver satisfaction.
- High specific output, bhp/litre/1,000 rpm (circa 17 + bhp) = high performance.

- Reduced valve overlap for a given output/engine capacity = low CO_2 and HC emissions.
- Lower compression ratio = lower peak combustion temperature = low Nox emissions.

Turbo Supercharging

During my time working as development engineer for David Morgan in the AML experimental department (later R&D) I had been involved in a number of engine projects involving turbocharging. In fact it was during a test run in the very first turbocharged V8 that David offered me a position within the experimental department. Rex Woodgate *(Mr Aston Martin in America)* had engineered the turbocharger installation with A.K. Miller in the USA and had sent the car over for evaluation. It certainly went very well with loads of mid-range torque. It didn't immediately come to anything in the way of a production version but was no doubt a pointer for the future. It was an influencing factor in the later twin turbo Lagonda prototype, which was probably the nearest that we came to an AML production version.

The first turbo development that I was involved with was the engine for the John Pope Special; this was also the most powerful at a whisker under 1,000 bhp on the test bed.

Much of what was learned from this engine was applied to the Bulldog or K9 project, which ultimately developed 728 bhp at 6,000 rpm on the test bed. Both engines were fuelled by the early V8 Bosch injection. We modified the metering unit internals at AML in the back room of the test bed building. The second-generation two-tone green bulldog prototype, the original car extensively reworked by a team led by Keith Martin, had the potential for a 237 mph top speed, which is what it was being prepared to demonstrate when it achieved 191.1 mph at MIRA. I was in the car

Figure 17.1 Showing the engine uncovered – no front cover on.

Figure 17.2 Engine from the front.

Figure 17.3 Right-hand side of the engine.

Figure 17.4 Showing left-hand side turbo and intercooler.

Figure 17.5 John Pope on the left and me on the right checking the engine at Silverstone.

with Keith when we did that run – he later joked that my sharp intake of breath as we approached the banked left turn after the timing straight could be heard in Nuneaton. You have to take into consideration that it was a left-hand drive car, which meant that I was up at the top of the banking with the steel restraining wires whistling past my right ear. To prove it, he did it again. Unfortunately, the car was sold before we completed the real high-speed run at NARDO and the project was abandoned. This was a shame, as Keith and his team had put a great deal of effort into getting the car right for that event, not to mention our efforts on the engine. A number of other manufacturers were subsequently involved in 200 mph publicity runs some considerable time later.

You can imagine that it was a major disappointment for us when the car was sold before we could do the high-speed run, particularly for Keith as it was intended that he would be the driver. But I think that the prospective new owner must have made us an offer too good to refuse and at the right time. But things didn't go too smoothly with that transaction. The evening of the day that the new owner had collected the car, we had a phone call to say that the engine had failed. Naturally the owner was not pleased and wanted the car fixed by tomorrow! So with the help of one of the experimental department technicians, Bob Clark, I rebuilt the engine overnight, pinching two cylinder heads from the endurance race engine project to speed things up. The engine appeared to have been seriously over-revved, with pistons and valves having had a coming together. After rebuilding the engine overnight, I took the car out on test the next day and all was well with the engine again, but on hard acceleration the clutch would slip, which explained why the engine had over-revved. The clutch was a ceramic racing clutch with loads of life left in it, but it had worn down and run out of adjustment. The customer wanted his car back *now*, so the only thing to do was to create more adjustment quickly! To achieve this we made up a 16 SWG shim to fit between the gearbox and the engine bell housing, which worked very well. So if anyone works on the Bulldog and wonders why the shim is there, you now know why. As a safeguard, we also fitted a rev limiter to the engine.

The original Bulldog had a turbo installation low down at each side of the engine, a repeat of the JPS engine. This configuration did not suit the Bulldog due to cooling problems associated with the mid-engine installation, so it tended to cook the turbos. The JPS car was front-engined and did not suffer to the same degree. The second-generation Bulldog was a much better car thanks to the work of Keith Martin and his team.

Within the engine department we had also extensively reworked the engine including mounting the turbochargers high at the rear on a pair of tuned length tubular manifolds. The air intake was via the lower slat covering the engine bay, the other slats were for rearward vision

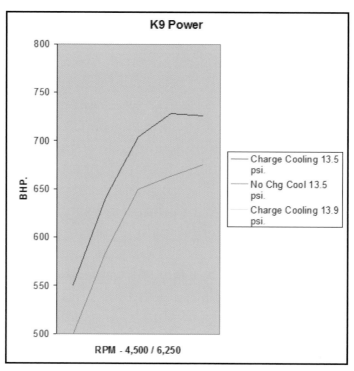

Figure 17.6 K9 Power Curve Sheet 1.

Figure 17.7 Bulldog and me after the road test. Note the big smile. I remember that engineering director Steve Coughlin made me *very* aware that it was costing £100 a day to insure the car for road test, a phenomenal amount in those days. So I had to be sure to let him know when I wasn't testing.

100

Figure 17.8 Left-side view of engine bay. Note the later position for the turbochargers with gauze covered air intake between the pair of T04s. This mated to an air filter mounted in the lower louvre on the engine cover.

Figure 17.9 Top of engine bay showing the special metering unit developed at AML. The metering unit boost regulator can be seen just in front of where the cold start capsule would normally be. Cold start enrichment was via a squirt of fuel into each manifold, operated by a small button on the dash.

Figure 17.10 Left-hand side showing wide door opening. Twin fuel tanks are mounted, one behind each seat.

Figure 17.11 Right-hand side view.

Figure 17.12 Eyeball to eyeball.

and air flow extract from the engine bay. It was this aspect of development that Keith and I had been testing when he made the 191.1 mph run.

The experimental Lagonda was probably the most normal turbo installation that we did. It was tuned for instant response at low engine speed by using two turbochargers as opposed to one larger unit; this was to reduce turbine inertia effect on response time. Also using relatively small *(for our engine size)* TO3 turbos further reduced turbine inertia and improved response. A large waste-gate was used to control boost and to prevent turbine over-speed. Bhp was not particularly high at around 375, but torque was very useful at a little over 480 lb/ft at 3,000 rpm. Coupled to the three-speed auto transmission, this provided startling performance for such a large car with no discernible lag in response to the throttle. At that time it was a blow through system via four modified Weber twin-choke 42 DCNF carburettors. Occasionally we would blow out the small lead gallery plugs in the carburettors. Petrol injection would have been used had we gone into production, which would have made it even better. Everybody who drove the car was impressed not only by the engine but by the way that the Lagonda chassis handled the extra performance. One journalist that I took out for a demonstration run commented that it was the nearest thing to a teleport system that he had ever been in. It could have been a great car.

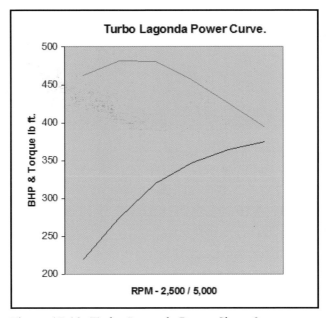

Figure 17.13 Turbo Lagonda Power Sheet 2.

We had an interesting time with the exhaust manifolds on this engine. We had a 5/16" thick manifold flange to which the tubular exhaust plumbing was welded. Because of the space limitations of the chassis the manifold was in the form of a log going forward to the turbos. What happened was that with the engine on full song, the manifolds became red hot and tried to expand in length. The 5/16" flange did not move, so the log manifold metal became compressed, it being at a workable temperature. When it cooled down, it contracted and no longer being at a workable temperature could not stretch back to the original length. The result was that it bowed the manifold flange so severely as to pull out the end-fixing studs from the cylinder head. To overcome this we had to cut the manifold into two parts joined by a slip ring joint to accommodate the expansion. If I remember correctly, I used a piston ring from an air-conditioning compressor for the slip ring joint. We would probably have used something like NiResist D5 for the manifolds had we gone into production.

There were a couple of marine engine projects, one for a 1,000 bhp race engine. We did a lot of work on this and spent many hours at the customer's boat yard but ultimately it didn't make it into production. We produced a mock-up section of the boat and learned a lot about offshore powerboat construction and engine installations. The engine was based on our experience with the K9 project and John Pope's 1,000 bhp race car engine. There was also a standard version, which did make it to prototype stage and was exhibited at the Earls Court boat show installed into a luxury high speed cruiser. AML at the boat show! This engine was based on the work that we did on the turbo Lagonda engine but benefited from being intercooled.

I had two other flirtations with the boating fraternity during my time at AML. The first was shortly after Bill Bannard took over as engineering director from Michael Bowler in the late 1980s. A well-established company that specialised in marine conversions wanted to do a marine version of our engine. They had seen the Tickford marine engine at the boat show and were quite impressed but wanted to do their own. Bill asked me to prepare a suitable specification engine and deliver it to the company for evaluation. I remember that it required a special sump and oil pick up because of the angle that the engine would be installed in the boat. The owner of the company was a very clever engineer; I was particularly impressed with his engine test equipment. Instead of the

Figure 17.14 Photo of marine engine showing the turbo installation.

usual dynamometer he had coupled up the engine to a large jet engine and applied load by shuttering down the outlet. The project progressed to in-boat testing but came to a halt due to a family bereavement.

The second time was when I was approached by a luxury boat-building company who wanted to resurrect that project. By that time we had moved on to the 32-valve engine so resurrecting the old 16-valve engine was not an option. However he became very excited when I showed him the torque curve of the supercharged Vantage engine. It was ideal for marine use with its massive torque from very low revs to get the boat up on to the plane. Also he would only need to install one engine as opposed to two, allowing for a narrower hull. From our point of view, I thought that this would be an opportunity to increase engine production numbers, which would reduce unit costs to AML. Although I doubt that it would have resulted in a massive initial order, it could have been a way into another market and you never know. I put forward a proposal but the AML directors of the time were not interested in pursuing the idea.

Returning to automobiles, the AM Tickford Capri was the only turbocharged engine project that made it into production. What an excellent car that was.

So we had enough experience of turbocharging to know that we could definitely produce the bhp required for the Vantage by that route, but there would be other concerns, particularly as this would be a catalyst installation. Catalyst light off from cold start would be expected to be slower due to the thermal mass of the turbocharger, which would normally be upstream. Heat handling under the bonnet would not be helped by having two hot masses in close proximity on each side of the engine: the turbo and the catalyst for each bank. Maintaining a short run to the catalysts would aggravate the packaging problems for the turbo position. Also I can remember that one of the problems with the under manifold position for the turbochargers on the original Bulldog was in providing adequate oil drainage from the turbos. Due to the low side profile of the 90 degree V8 engine configuration, the turbochargers were not much above the engine sump oil level. This could cause drainage problems particularly under high G forces. This was considered to be one of the main reasons for the high turbo failure rate during early track development. Another thing that we learned from that project was that turbos do not like backpressure, especially on the higher output engines, so we would have to expect that the effect of catalyst backpressure would reduce the effectiveness of the turbocharging. Typically at that time, we would expect to lose 60 bhp from a normally aspirated engine due to the effect of the catalyst. Not just in terms of back pressure but also because of the constraints on optimised primary lengths. Previous data would suggest that a 100 bhp drop due to backpressure would be realistic for a turbo version at the intended output of around 500 bhp. It is interesting to note that turbocharged petrol engines were not as popular for a while after the introduction of catalysts. Nevertheless turbocharging was a way of achieving our objectives. And it must be said that catalysts have improved dramatically over the last ten years since we started the Vantage.

Exhaust Gas-Driven Turbocharging

Advantages
We had done it before.
Low torque at low engine speed.
Bhp target could be met.
Impression of performance is greater than actual.

Disadvantages
Very hot parts under bonnet.
Turbo position and height limitations imposed by oil drain requirements.
Adequate oil drain required. Could require separate scavenge pumps.
Driver education, i.e., shutdown procedure.
Thermal inertia effect on catalyst light off.
Catalyst backpressure effect on performance. Typically 80/100 bhp loss dependent on tune.
Pressure oil supply required.
High performance exhaust gas sealing required – manifold, gaskets, etc.
Use of specialised materials for exhaust manifolds.
Engine pumping losses.
Turbo lag effect on driveability.
Exhaust manifold installation/wide 90 degree V8 engine.

While I was working for David Morgan we had looked at engine driven supercharging on a number of occasions. But it was difficult to find superchargers of the right capacity. What were available were mainly from the after-market with no proven reliability as original equipment. Supercharging tended to be an after-market activity very much as add-on equipment. One of our contacts in particular had demonstrated a very neat installation albeit on a smaller engine. He visited me shortly before the Vantage engine project to demonstrate a V6 Capri that his company had converted. It went very well but didn't resolve any of our particular concerns at that time. Nevertheless it was a timely reminder of the smooth power delivery available from a supercharged engine.

There had been one other attempt at supercharging a Lagonda – this was a project that Bill Bannard had agreed to another company doing. It was at the time when we were in the midst of the 32-valve engine development and Bill had warned me off of the project as he needed me to put all my effort into the 32-valve job, which was a shame as it showed great promise. The installation of two neat little belt-driven centrifugal superchargers was very well executed. The car went extremely well if you could plug your ears to the detonation rattle – it just needed an engine guy to get to grips with the base engine and the calibration. Needless to say, it didn't go into production.

At the time that I was thinking seriously about the Vantage, Richard Long of Eaton contacted me to tell me about a Roots type supercharger that they had developed. It was being used by Ford on their new Thunderbird. He could not have timed it better. The Ford connection was particularly appropriate in the light of subsequent events.

Power Delivery

The two types of supercharging are quite different in the method of power delivery. An engine-driven supercharged engine would have increased power over the

Engine Driven Supercharging

Advantages
Well-proven parts – Ford product validation test data available.
Self contained lubrication system (no oil drain back to engine sump).
No effect on catalyst light off.
Catalyst has less effect on output, typically 20 bhp loss.
No special driver instruction required.
Bhp target achievable.
Instant response throughout the engine operating range.
Power increased over the full operating range of the engine.
High torque to pull high gear at low engine rpm during cruise giving high mpg.
Off boost losses could be minimised by a bypass configuration.

Disadvantages
Belt drive system.
Unit size and weight.
Driven losses.

full range of operation, steam engine like torque from idle to max power. A turbocharged engine would tend to come on song as engine speed and load increased or in transient conditions with a degree of lag in response followed by a swell of delivery as things start to happen. The latter is possibly more exciting and to some extent flatters the actual performance. From previous experience, we knew that if the complications associated with catalysts could be ignored, it would have a greater potential for out-and-out bhp. A compound system combining the characteristics of both forms of supercharging could be the ultimate.

Engine Design Implications

The engine design priorities are quite different for turbocharging as opposed to an engine-driven supercharger.

The turbo is dependent on exhaust gas energy to drive the turbine so the better the engine breathes naturally the more energy is available to drive the compressor turbine. Thus the turbo is dependent on the state of tune of the base engine especially so at higher outputs. Whereas the engine driven supercharger is less so raising the possibility of more modest valve timing to control emissions.

Callaway were making good progress with the racing engine for AMR1 and there was a strong leaning from within the company towards having Callaway design the Vantage engine. They had also been involved in a number of successful turbo-charged applications of their own. Andrew Woolner, who was engineering director at the time, and I visited Callaway Engineering for preliminary discussions and, as before, they were very supportive and flexible in their approach. They favoured the turbo route but considered our target output a significant challenge. But by this time I had gained some support for an engine-driven supercharger application and it was eventually decided that we would go via this route as an in house engine project.

My recommendation was that the engine for the Vantage be based on a twin-supercharged configuration. But that an immediate investigation into the then current cylinder head should precede supercharged engine development. This recommendation was accepted by the engineering director, Mr Woolner, and the chairman, Mr Gauntlet.

Supercharger Installation

The performance data for the Eaton M90 supercharger showed that we would need two units for our engine. And that with a 1.8/1 drive ratio, it would take around 30 bhp to drive each one at 6,500 rpm engine speed on full power: 30 bhp that we would not otherwise have and that needs to be weighed against the increased pumping losses associated with a turbo, nothing is free. Eaton conservatively predicted an output of 460 bhp at 6,000 rpm. We intended to use a drive ratio of 1.8 to 1, which complied comfortably with the recommended maximum continuous supercharger speed of 12,000–13,000 rpm.

The complexity of the drive configuration for two units would be offset by the better packaging opportunities of two smaller units, rather than a single large unit. Also we would be able to keep the induction volume down to a minimum by treating each side of the engine separately.

Richard Long supplied us with two slave M90 superchargers with short nose housings to mock-up an installation on a slave engine. The first mock-up was in a Lagonda chassis as at this time it was intended that the William Towns-styled Lagonda would also use a version of this engine – wouldn't that have been great? There really were only two installation options for a twin system, one being on top of each cam cover, which is where they are. The other was close to each side at the front of the block low down under the exhaust manifolds. This was always the least attractive option, but as always bonnet line height would be critical.

> It is very sad that the Lagonda name appears to have been forgotten. And that we didn't go ahead and fit the supercharged engine into it. It would have made a wonderful combination of luxury and effortless speed.

A mock-up of each supercharger position was made to identify the areas of concern. The cam cover-mounted position was always the favourite. The only real problem would be to get the air charge from what would be the bottom side of the supercharger out and along the side of the engine where it would be very close to the chassis, and then into the intercooler.

It was accepted that the adiabatic performance of a Roots-type supercharger would not be as good as that of a centrifugal compressor of a turbo. So it was always intended that the engine would be intercooled. Although, to be fair, a turbo version would probably need to be intercooled to maintain the intended output, it would certainly be better for it. There were plans for top mounted air-to-air intercoolers with NACA ducts in the bonnet as I had done on the Zagato Coupe, but the styling department didn't like the idea. The alternative was to run an air to water system with remote water radiators. The wet intercoolers would be fairly bulky so would need to be mounted in front of the engine en route to front entry plenum chambers. All pipe work is in cast LM25 with rigid mechanical connections to avoid noise being generated within the system. Eaton had warned us that this would be essential.

A centrally mounted crossover inlet manifold would accommodate two plenum chambers in a good position to

Figure 17.15 Early mock-up showing supercharger outlet connection to the intercooler.

provide a short bypass route back to the inlet side of each supercharger. This bypass arrangement would reduce the drive losses during off-boost conditions by virtue of the reduced density of the air charge being circulated by the superchargers. The bypass route is back into the inlet volume, downstream of the throttle. The recirculated low-density air charge would also be cooled by the inter-cooler in the process, which in turn would reduce any residual heat in the supercharger body.

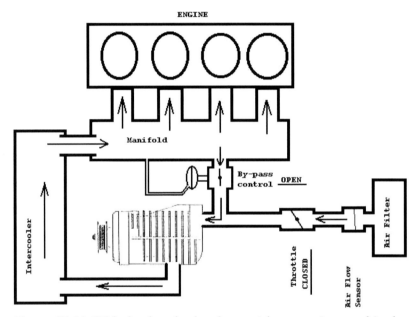

Figure 17.16 With the throttle closed, a partial vacuum is created in the manifold by the operation of the engine. This vacuum sets the bypass control to open reducing the energy required to drive the supercharger by virtue of the low density of the air being moved by the supercharger. By recircu-lating the air through the intercooler any residual heat retained by the induction system and the supercharger body is reduced.

The inlet manifold is cast in one piece with inlet runner lengths modelled on the length optimised on the nor-mally aspirated engine. The left-hand supercharger system feeds the right-hand bank of the engine and vice versa for the left bank. Each side of the engine has its own separate intercooler system and fuel system.

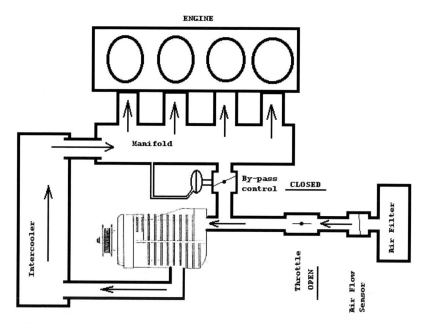

Figure 17.17 As demand is increased to the engine by opening the throttle, the vacuum within the manifold decreases to a point where it is unable to hold the bypass control open. As the bypass is closed, the full effect of the supercharging is delivered to the engine in a smooth and seamless fashion.

Engine Management

In March of 1990 I visited the Weber factory at Bologna with Graham Bangs from Weber Concessionaires in Sunbury *(now Webcon)*, the intention being to persuade our friends of many years at Sunbury and Bologna to support the Vantage programme and to discuss my shopping list of engine management and fuel injection equipment. This included progressive twin throttle bodies for each side of the engine, coil on plug to avoid having HT leads, high delivery *(circa 48 gallons/hour)* fuel pump and high-pressure regulator to respond to boost pressure. Not to mention injectors with high resolution at low delivery for good emission control and sufficient flow to feed 500-odd horsepower. There was a real possibility that twin injectors per cylinder would be required to cover the range of operation.

The Weber engineers were enthusiastic about the project but were not keen on the idea of the supercharger being downstream of the throttle. This was a fundamental feature of our design and of the bypass arrangement to reduce off-boost losses and control fuel consumption. There were long discussions about this and eventually they were convinced to the point where they would start the project as long as their concerns were noted. Later on in the project they agreed that we had taken the right decision, but not before their concerns had reached Mr Woolner's ears and he had insisted that a downstream throttle system be designed and evaluated. It was not what we wanted – it was a plumber's nightmare and the superchargers would need to be disengaged by electric clutch. Happily this did not progress beyond the mock-up and drawing stage before the penny dropped.

Of the shopping list, the coil on plug units were too big to fit into our cylinder heads and the fuel delivery would need to be via two pumps. Other parts were no problem including the injectors, which had been a concern, although things would start to dry up just above 600 bhp with single injectors. While there, I was shown the latest in fuel rails and injector harness conduit equipment. There was nothing there that we could use straight off the shelf but it prompted a look into the possibilities of cast aluminium fuel rails for the Vantage. The picture of the mock-up in Figure 18.1 shows the Virage type fabricated fuel rails.

Figure 18.1 The Weber system, ignition distributors driven from the front of each inlet cam. Note air-to-air intercoolers fitted for mock-up only.

Almost all of the base engine development was completed on the Weber injection system before we were asked to reconfigure the engine to use the Ford EEC4 system manufactured by our new parent company. This was a major design change as the EEC4 system is a mass air flow (MAF) system, requiring air flow meters to be installed into the induction. This introduced two major

Figure 18.2 Rear view of early development engine with Weber injection. The bypass valve can clearly be seen just forward of the left throttle body going into the plenum chamber. As can the crossover inlet manifold casting.

problems. One was that our induction system was designed as two separate sides of the engine, so would need two flow meters or a complete redesign of the induction system. Also the engine management strategy for two MAF sensors did not exist at that time. So Steve Armitstead, our new electronics genius, had to get to work on that. True to form, he came up with an ingenious strategy to solve the problem for us. This only left the other problem of how to package each air flow meter into a position upstream of the relevant throttle body. In the Weber configuration, the throttle bodies were rear-facing with air filters at the back of the engine drawing the intake air from the bottom of the windscreen. So they were already very close to the back of the engine bay. It is usual to site the MAF sensor in a straight part of the induction to avoid turbulence in the airflow through the sensor. We didn't have room to extend the intake to include a straight section. Our solution was to install the air flow meter into the air filter housing using the position of the back of the housing to correct the flow to the engine. This design was based on some work that we had conducted during the endurance race engine development to overcome an induction height and body panel conflict. By running a plate positioned exactly 1" above the trumpets, we could achieve the same power from a shorter than ideal intake trumpet. The plate could be part of the air box enclosure. In the case of the Vantage it was the back of the air filter housing.

This variation on a theme worked well on the Vantage engine and the flow characteristics were later confirmed during the air flow meter calibration work conducted by Steve and the Ford people in USA. The air filter housings on the Vantage look pretty ordinary but they are a good example of effective packaging.

I have to say that the engine made slightly better power when using the Weber engine management system, possibly due to the complications introduced by the airflow metering hardware of the EEC system. However due to the delays in releasing the Vantage the on-board diagnostic (OBD) requirement was beginning to loom in the not too distant future. At that time, it was considered that the mass air flow measuring approach adopted by the EEC system had a greater capability of meeting that requirement, which in itself was still pretty hazy as the OBD rules were still being formulated. That and the obvious benefits of us using a Ford product were the only reasons for the change.

Base Engine Preparation

Having decided that supercharging was the way that we would go, we would need to ensure that the engine would survive an increase in power output of the order of 80 per cent. A significant increase in brake mean effective pressure (BMEP) would be expected which we knew the Virage engine would not tolerate. We needed to continue our investigation into what was happening in the combustion chamber.

Also an 80 per cent increase in power would increase the loading input to the engine structure. Bearing loads and effects on block stiffness would need to be thoroughly studied. Cylinder pressures were expected to exceed 110 Bar. A Formula One engine of the time ran to 100 Bar, so cylinder head sealing would also be a subject for study. The Vantage engine would have to be designed as a high-output supercharged engine from the outset. Such an increase in power would mean that this could not be an add-on exercise and it would be a lot more involved than just bolting on a couple of superchargers.

Also, from a historic point of view, this would be the first Vantage engine not to be based on the standard version of the time, which was the Virage. In fact the process would be reversed; the next normally aspirated standard engine would be partly developed as the basis for the Vantage project. It would be released some considerable time after the Vantage as the V8 Coupe engine.

Cylinder Head Development
To mount an in-depth evaluation of the cylinder head would require the use of specialised test equipment

not available at AML, so our old friends at Tickford Engineering were engaged to carry out this work on a Virage engine that we had specially prepared with the necessary combustion chamber, pressure sensors, etc.

As a result of that evaluation, Tickford were commissioned to redesign the inlet ports to reduce the mean gas velocity and to promote tumble in the combustion chamber. The exhaust port was also to be redesigned to improve gas flow and ultimately to incorporate a new air injection port to each valve. New valves and guides were designed as part of the exercise to improve gas flow. The improvement to the exhaust port was in part made possible by improvements in casting techniques, which made the rather intrusive air gallery unnecessary. And, to be fair, the original Virage inlet port had been compromised from the outset by having to make it round at the manifold face to match the old manifolds from the two-valve engine, which we didn't use after all.

The results from the Tickford work provided a marked improvement. As well as increasing tumble or barrel swirl by 62 per cent, the new ports and valves also flowed better. To some degree, flow has to be sacrificed in the interest of creating sufficient tumble so a balance has to be arrived at. In this case both were improved. Water circulation around the spark plug boss was improved by the lower position of the new inlet and exhaust ports. It also gave a longer support and heat path to the valve guides with less guide intruding into the ports. As a result of this work, the engine was no longer detonation limited so optimisation of fuelling and ignition timing was possible throughout the speed range. Maximum torque was improved by 5.5 per cent and maximum power was improved by 9 per cent and ignition advance requirement was reduced by up to five degrees, indicating improved combustion.

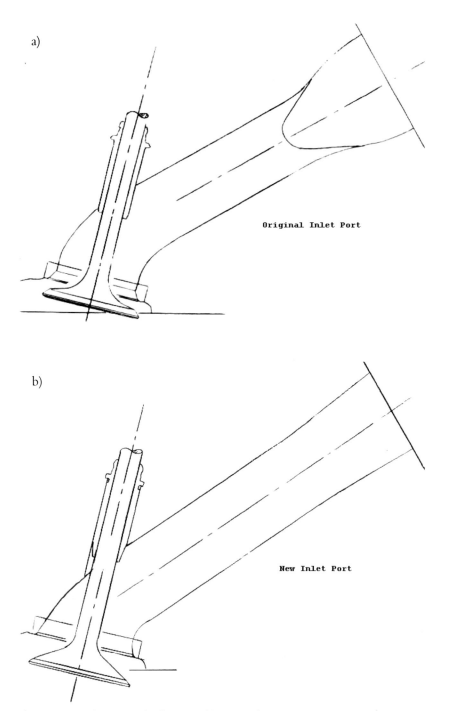

a)

Original Inlet Port

b)

New Inlet Port

Figure 19.1 a) Original inlet port; b) new inlet port.

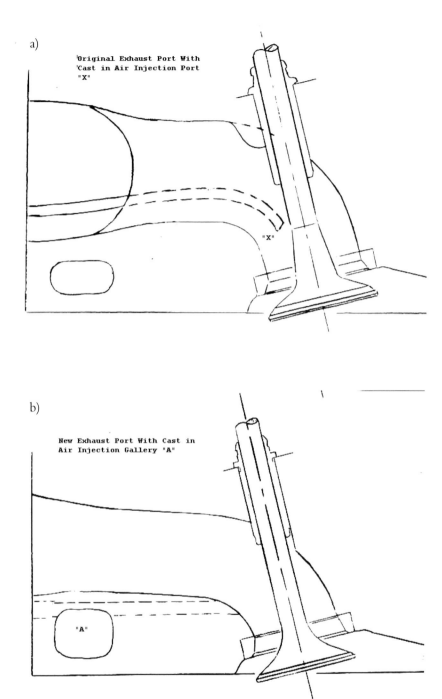

Figure 19.2 a) Original exhaust port; b) new exhaust port.

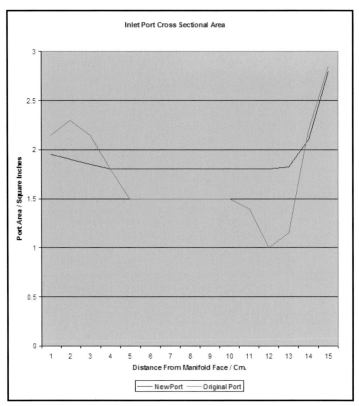

Figure 19.3 Inlet port cross sectional area. Port area.

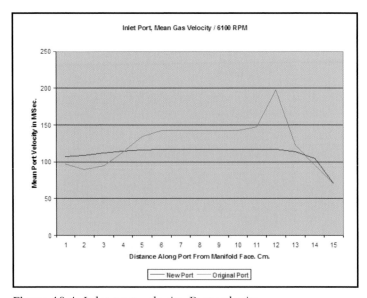

Figure 19.4 Inlet port velocity. Port velocity.

Combustion Chamber

During the above work the opportunity was taken to have Tickford evaluate the combustion chamber redesign that I had done early in the four-valve development. Tests were conducted with and without the between valve squish areas but no improvement could be made so the chamber was retained as it was.

Cylinder Head Casting

Cosworth were now casting the heads for us as Birmal had closed down. George Wright and his team at Cosworth Castings were very much involved in the new design and were keen to exploit the casting technology of the time as well as making the changes resulting from the Tickford work. A cast in oil gallery complete with the oil ports to the hydraulic tappet bores was made possible. Positioning of the oil ports is critical to the correct function of the hydraulic tappets, so this was a major move forward in casting precision. Previously this had involved a complex drilling operation to ensure the required accuracy. Oil drainage was improved by the introduction of a drain from the rear of the cylinder head, which was matched to the proposed new cylinder block, the connection being via a bobbin-type transfer port. The check valve housing for the left-hand bank cylinder head chain tensioner was deleted and the check valve moved into a new compact bronze tensioner housing. This was to reduce the volume of oil between the check valve and the tensioner plunger to combat an oil aeration problem at high oil temperatures. At the same time, the head casting was generally cleaned up by the removal of some unused features. Wall thickness was reduced where possible to maximise coolant flow and to reduce weight.

We had experienced problems with the quality of the special 5/16" UNC studs used for the Virage cam bearing caps and the cam cover fixings on the engine. The cam cover nuts were also unique and involved O ring seals. So the cam cover face was changed to accept a simpler, more traditional, Aston cam cover design using standard fixings that did not require seals and were independent of the cam bearing fixings. This meant that the cam bearing caps could also use standard fixings independent of the cam cover. A side-effect of these changes was a significant reduction in costs and build time. To avoid any colour match problems, all covers were powder-coated to match the colour of the Eaton superchargers, which were fortunately quite close to our traditional V8 powder coat colour if slightly darker. This provided better quality control of colour, as opposed to the anodised finish on the Virage engine.

Cylinder Head Durability Testing

Durability testing of the new cylinder head with new ports and valve gear began back at AML. Initially all testing would be in normally aspirated form prior to the first supercharged engine being built. The test cycle used was based on the British Leyland test BLS 53.06 and would run for 200 hours on bed 1. This testing identified a valve seat retention concern. Two engines were damaged at the same point of the test cycle before we were able to establish the cause of failure. The damage to the valve and valve seat area was very extensive, which did not leave much in the way of clues. But by looking at the part of the test cycle that caused the failure, we could see that it involved a sudden change from high output at high rpm to a light load high rpm condition. We considered that this had caused rapid cooling of the combustion chamber casting, causing contraction onto the valve seat inserts that were still at very high temperature, resulting in compression of the seat insert, which then became loose as it cooled.

To confirm this theory, a two-dimensional computer study was conducted to investigate the thermal stress input to the valve seat. This confirmed that the valve seat could yield due to compressive hoop stresses induced by the high duty thermal load cycle. The valve seat material has a yield strength of the order of 1,170 mpa at 300 degrees Celsius. As a result of the analysis, the valve seat interference fit was reduced from 0.0035/0.0045" to 0.0015/0.003" and a further 200 hours of the durability cycle were completed without further incident.

Coolant Flow

We had conducted an extensive temperature survey of the standard Virage engine on the AML test bed, which showed that the third cylinder from the front on each bank could run slightly hotter than the other cylinders if output was increased. This situation was further complicated by the heater take off from each bank, which could influence cooling dependent on heater demand. This tendency was not peculiar to the Virage engine; it had also been an area that had been the subject of attention on the 16-valve engines at various times. The extra heat that would be generated by the Vantage engine would make this something that we would need to look at, particularly during the design of the new cylinder block.

To observe the coolant flow, we machined through the top section of a cylinder head and sealed the section with a thick Perspex sheet, with the test head fitted to an engine block casting. We hoped to study the flow route by pumping water through the engine and introducing air bubbles or dye into the flow. This was only partially successful. However, I knew that Richard Sykes at Tickford had had some success in a similar situation with a high-speed camera, so I spoke to Richard and he agreed to try. Unfortunately the camera idea didn't work either, but by trial and error we eventually felt confident that we could interpret the flow through the engine. By adjusting the transfer port sizes and repositioning the heater take-off, a more balanced flow through the cylinder head was achieved. Later on we were able to repeat the temperature survey on a full Vantage spec engine on the test bed. And with a few minor adjustments, we achieved an even temperature balance from front to back of the engine.

For a short time a special limited edition of the Virage was produced, which used the new Vantage cylinder head. This was the first normally aspirated engine to use the new head. It was equipped with appropriately modified (oval port) inlet and exhaust manifolds, but otherwise the engine was standard Virage including the engine management calibration. Luckily we had left the fuelling slightly rich for the Virage to help with the detonation problem. Power was immediately increased by 6.5 per cent to 330 bhp. The standard Virage continued in production with the original cylinder head. It is worth comparing this power curve (Figure 19.5) with the original Virage power curve. Note the torque curve has no dip in the middle.

Cylinder Block Changes

Birmal had also been the cylinder block casting supplier at the time and, as has been said, were closing down, so a new source was required. For this we first went to Zeus castings and then to Grainger & Worrall. The opportunity was taken to specify a material change to LM25, this being more widely used for this application than LM8 due to a reduced tendency to porosity or blowholes. Pattern work was completely reworked to encompass two different cylinder blocks – the old 16-valve V8 engine and the new Vantage. The Virage engine would use the Vantage block but with blanking plugs fitted to the rear oil drain cores. Changes for the Vantage block included reduced wall thickness and increased waterways for improved cooling,

128

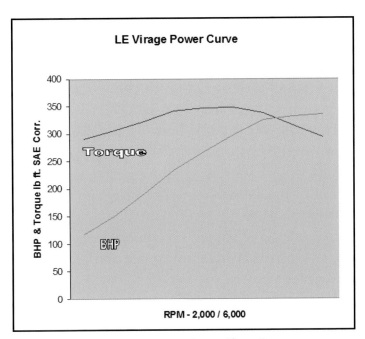

Figure 19.5   LE Virage Power Curve Sheet 3.

particularly around the rear cylinders; our flow work had identified a restriction around the back of the rear cylinder liners as well as a new rear main bearing housing that did not include the old scroll-type oil seal. Pattern work was made to accommodate a choice of rear main bearing seal for the new or old type to support service requirements. The new bolt-on housing incorporating a lip-type seal for the rear main bearing would provide added cross-brace support to the main bearing cap buttresses at the back of the block. We would also make provision for drain channels from each cylinder head at the rear of the block. To further strengthen the block, we would also de-sign out some of the machining operations. This was to avoid stress risers, particularly in the rib sections at the sides of the block near the sump face. Previously the spot face operation for the sump bolt positions had resulted in the ribs being machined through. To avoid this, the spot faces for sump bolts and the water drain bosses were cast to a finished dimension. The drawings were also changed to move the main bearing cap positioning datum from the left-side stud position, to the crank centre line. This was to centralise any tolerance effects and allow us use a closer fit for the bearing caps into the block. Bearing cap clearance fit into the block is now 0.001" +/− 0.0005", and on later engines the side buttresses in the block casting have been extended down to the sump face. The main purpose of this change is to allow better venting during the casting process, improving the integrity of the casting around the main bearings. It also adds to the stiffness of the block casting. All Vantage block castings were X-rayed for integrity prior to machining, particularly around the main bearing areas.

The deep-skirted block was also given additional support by using an anaerobic seal instead of a conventional gasket on the sump joint to effectively glue the two parts together. The old Thackeray-type spring washers fitted to the sump bolts were also changed to serrated lock washers to provide a more positive clamp load. The new rear oil seal housing is designed to provide a bridge support at the rear main bearing position replicating the support provided by the oil pump at the front of the block. The fixings for the internal baffle plate at the base of the block had already been upgraded to Durlock, which also enhanced the rigidity. The main bearing caps were upgraded to forged RR58WP Dural for the Vantage, previously they had been in RR56WP.

Cross-bolted main bearing caps were considered. But my calculations showed that the moves to stiffen the block, combined with the tighter limits on the main bearing cap fit, would be sufficient for the intended maximum engine speed of 6,500 rpm.

So there were some major changes and quite a few detail changes that are not immediately obvious unless you know where to look, but the resulting cylinder block is structurally much stronger and stiffer with increased waterways for better cooling. Thermal stability was also improved, we conducted cylinder bore ovality checks over the full operating temperature range for the cylinder block and found a maximum of 0.001" ovality for any one cylinder bore over the whole range. This was particularly important to us, because a few years earlier, when we were looking into the effects of bore distortion on oil consumption, we had experimented with fitting the cylinder liners to a tighter fit into the block only to find that bore distortion worsened. Indications were that the cylinder block casting was moving more than the cylinder liners were during the heat cycle.

Holding the tight tolerance for the bearing cap fit would require a high degree of accuracy on the part of our machine shop. But Des Lovett, who was in charge of the machine shop at that time, was confident and he didn't let me down.

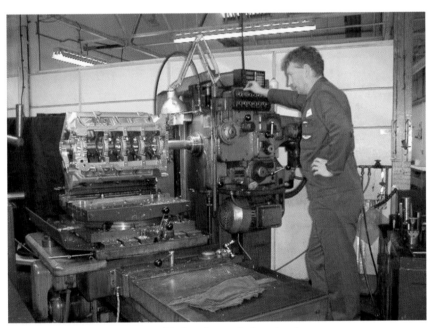

Figure 19.6 Des Lovett machining the main bearing bore.

Figure 19.7 Line boring the main bearings. The extended side buttresses to the main bearing and the new rear main bearing cap can clearly be seen in this photo.

Crankshaft

The crankshaft forging was altered to extend the material around the throws. This was to improve the side thrust support to the connecting rods for the full diameter of the crank pins. To compensate for the added material on the crank pin shoulders meant that the crankshaft forging counterweights had also to be increased slightly. The rear scroll and flywheel flange was redesigned to form a new lip seal surface incorporated onto the outer diameter of a 100 mm diameter flywheel flange. Surface speed limitations for the one piece seal dictated the size of the flywheel flange. The flywheel flange fixing had to be incorporated within the 100 mm diameter of the new lip seal to allow for fitting of the seal and housing. This meant that new flywheel fixings were required to take the extra load generated by them being at a much reduced pitch diameter. In this area, some early cranks had two sets of fixing holes, one set being to design protect for a dual mass flywheel during development. The standard flywheel at the time weighed 19.5 lb, but this had to be increased to 24.25 lb to enhance the idle quality of the Vantage. The dual mass flywheel weighed in at 55 lb and was not considered appropriate for our application so was not used.

Figure 19.8 New crankshaft showing the rear seal land, counterweights and increased material around the big end pins.

Crankshaft Bearing Loads

A computer program analysis of the predicted bearing loads identified a slight concern regarding the standard bearing set-up, nevertheless our experience at higher outputs gave confidence to retain the Vandervell VP2 bearing shells. These had been used in the endurance race engines of 1982 onwards and had initially been introduced into production for the 16-valve Vantage, a slightly softer Glacier bearing having been used previously. In response to the analysis it was decided to provide greater support to the bearing shells by reducing the housing diameter by

0.0005". Beyond this action, we would fit plain, non-grooved shells to the lower half of the bearing should the need arise, but it didn't.

Conn Rods

Connecting rod forgings had been beefed up for the endurance race engines and those forging changes had been carried over into the standard production engines. Thus the rods for the Vantage are from the endurance race engine forgings but they do not use the spiral pins for cap location and the standard 15/16" gudgeon pins are retained as opposed to the 1" pins used in the endurance race engine.

Pistons

The Cosworth-forged pistons were originally solid skirt but were changed during development to include a thermal barrier slot to reduce engine noise. Two new features were also introduced into their design: diamond-turned skirts for scuff resistance and lubrication retention and scuff bands above the top ring land. These allowed for a closer running clearance and reduced the HC trap usually associated with that area as they fill up with carbon. The pistons were forged in RR 58 TF.

A low friction ring pack was used with 1.5mm wide compression rings. This again was an effort to reduce friction-generated heat and in response to the high frictional losses noted during our earlier engine evaluation.

Controlling frictional losses was important to the Vantage project both in terms of power loss and probably more importantly in terms of heat input to critical components like pistons. The small end bearing clearance was increased to 0.001" to allow for pin expansion during hard running, avoiding the pin locking in the rod and as a result rotating in the piston, which would then generate heat back into the piston boss and thereafter into the crown. If you think about it, the heat path is piston crown, pin bosses, pin and then small end bush, so an allowance has to be made for the pin to expand in the bush. I remember that the criteria for the small end fit on the six cylinder engines was a thumb push fit for the pin at room temperature. How times have changed.

The low friction ring pack was important to the Vantage for the same reasons. But despite extensive testing during our development trials, some customers experienced unacceptable rates of oil consumption. We eventually traced this to an overrun condition whereby the engine oil was going around the back of the compression rings during lift-off. The cure was to incorporate a small step into the top land of the second ring grove to induce ring twist during overrun conditions. Not sure why we didn't find it before we went into production, as we were getting 800 miles to the pint regularly. But maybe that was down to the small test sample that we had or maybe our test drivers didn't do much in the way of lifting off. Testing for overrun oil consumption is not easy without access to a special overrun dynamometer, which we didn't have. The only way that we could do it was to devise a specific test drive cycle to be conducted by our test drivers on Millbrook test circuit.

Gaskets

We were about to change all gaskets to non-asbestos composition to comply with new regulations so this was done as part of the Vantage project. Two types of cylinder head gaskets were developed based on clamp load tests using Fuji Prescale film and an eyelet configuration developed during the endurance race engine days. The more even stud spacing of the 32-valve engine gave a very uniform clamp load. We also tried a Coopers ring type gasket with specially machined liners, this was generally considered as the brick outhouse solution should all else fail. But true to Murphy's Law, it failed. It was not pursued further due to successful testing of the more conventional eyelet type gasket with screen print seals for water and oil ways. This gasket has proved to be very reliable.

Supercharger System Belt Drive

The Eaton M90 supercharger is a short nosed version of one used by Ford US on the Thunderbird and is a Ford-approved and tested part. The same applies to the bypass control actuator valve and the drive belt tensioner units. Although the latter have had to be recalibrated in line with the belt slip and system loading test work conducted in house. This testing was done in collaboration with Litens Automotive, who manufacture the belt tensioners.

The significance of having Ford-approved parts was that it would make it unnecessary for us to do our own product validation testing. This type of testing can be a lengthy and costly process.

Figure 19.9 The belt drive system of the Vantage engine. Intercoolers are removed for clarity.

Figure 19.10 View from above showing the right-side supercharger drive forward of the left side. The right hand supercharger serves the left bank of the engine, which has its cylinders offset by 0.938" forward of the right bank.

Catalysts

Catalysts were always going to be a problem, back pressure would be critical, as would light-off in a position far enough downstream from the exhaust valve to allow for a reasonable exhaust primary pipe configuration. Also to be protected from reaching excessive temperatures by being too close to the by-products of 500+ bhp. A matter of being hot enough at light load and not too hot at high load, the difference in this case could be considerable. Over the years we had developed a good understanding of lighting off catalyst that by most standards would be considered too far from the action. But this would be a tough one and we knew that this knowledge would be tested to the full and, to that end, air injection would be used during start up to aid light-off.

At first there didn't appear to be a catalyst available that would do what we wanted. It looked like we would have to develop something special for the Vantage. A design for a catalyst incorporating a flap valve to divert from a start-up catalyst was played around with, but the cost of such a device was going to be very high. I think that Ferrari had dabbled in this area at some time. Johnson Matthey was very helpful and eventually steered us in the direction of what had been a joint project between Porsche and Eberspacher. This was to develop a metal substrate catalyst specifically with low back pressure for the Porsche 911 Carrera 4, the development being documented in SAE paper 890488. An agreement was reached with Porsche and Eberspacher to supply us with a variation of the catalyst. The changes for our use are to entry and exit flanges, which would be done as part of the exhaust system manufacture and to the precious metal loading, which Johnson Matthey would do for us, 60 grams/ft^3 instead of 50 as used by Porsche. None of which was judged to compromise the durability testing that had been conducted by Porsche. Their testing had included the emission conversion durability, which was required for the USA, so we wouldn't have to repeat that test, which would be a major saving. Using two of these units would give us a catalyst ratio of 0.9 of engine capacity.

We committed to this route and contacted Mike Marriott at CLF International to fabricate a manifold that would fit into the car with the primary lengths that we needed. Initially we thought that we may have to settle for a two into one, two into one configuration and accordingly a design was made as back-up. But CLF came up with the goods and gave us our first choice an excellent 1¾" diameter four into one manifold that they squeezed into the very small space available. Primary lengths were compromised again in anticipation of cat light-off problems but all primary lengths were the same at 30". A test bed set was then made to the same configuration as the ones that fitted into the car, but with extendible primaries for development. Cold start and low-speed catalyst light off optimisation was achieved without exceeding a maximum mid bed catalyst temperature of 1,000 degrees Celsius at sustained full power. Catalyst durability has been excellent with only one pair of catalysts being used for all of the original test bed development and durability testing.

Figure 19.11 CLF International – exhaust manifold.

Spark Plugs

I have known Jim Hughes for a long time as a friend and as a director of NGK. He is a great enthusiast and has always provided superb support to AML. Martin Soles was technical services manager for NGK and it was Martin who would support the Vantage project personally. We have had a longstanding relationship with NGK spark plugs and they have certainly got us out of a couple of sticky situations in the past. This relationship has developed a high regard for the product and for the level of technical support provided. There was no doubt in our mind that the wide heat range offered by NGK spark plugs would be an absolute necessity for our application. The temperature range demands placed on the spark plugs during the change from low speed traffic driving; to full out 550/600 bhp operation would be extreme.

A spark plug thermocouple test on the AML test bed showed that seven heat range plugs were just a bit borderline so we played safe and specified an eight heat range plug BCR8ES. Test bed and road test durability mileage with the BCR8ES grade plugs indicated that 24,000 mile replacement intervals were just feasible. But the recommended service interval is 12,000 miles, bearing in mind the difficulty of changing plugs at the roadside and the costly effect of misfire on catalyst life and the high performance nature of the car. Precious metal electrode spark plugs would only be specified to comply with the 30,000-mile federal requirement.

> The superchargers have to be removed to change the spark plugs, a fact that I have been constantly ribbed about, by our service department technicians. All in the best possible humour, I think.

Camshaft

Initial development started with the 0.4" lift Virage cams, which comfortably achieved the target bhp output. But torque was higher than we wanted, with more than 600 lb/ft easily available. This gave concern for transmission reliability and was well outside the recommended maximum for the gearbox and axle, particularly as the curve was very flat with more than 500 lb/ft available at 1,200 rpm. We could back off on the supercharger drive ratio or tilt the power curve towards more bhp and less torque by running more valve overlap. Which is what we did we ran a new profile 0.415" lift camshaft with a 268 degree period in both sides again. This gave a more balanced output with torque and bhp both around 550. But with more than 85 per cent of full torque available from 1,500 rpm upwards, the driveline was still in for a hard time. When designing the new camshaft, the opportunity was taken to introduce a cutaway base circle to aid tappet rotation. The earlier cam had a tapered nose profile to encourage rotation. This occurred at full lift when loads are high. The cutaway base circle induces tappet rotation when the valves are closed and loads are at their lowest.

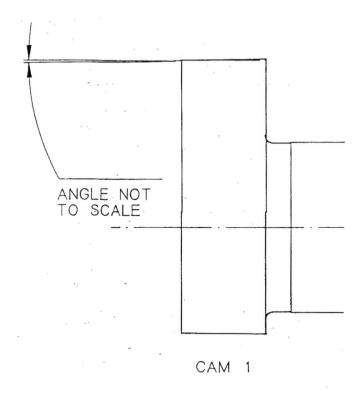

ANGLE NOT
TO SCALE

CAM 1

Figure 19.12 Taper nose cam.

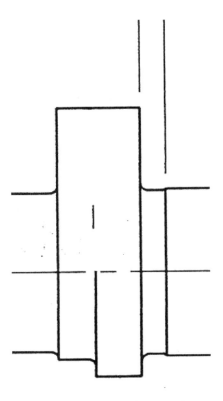

Figure 19.13 Cutaway base circle cam.

Our new association with Ford allowed us access to their camshaft design facility. We asked them to give our cam profiles the once over with a view to improving cam and valve gear durability which they did. Small but subtle changes were implemented that resulted in a more refined operation.

Lubrication

Because of the high performance potential of the car, dry sump lubrication was considered. It would improve the power output and control lubrication at high G force, and we had experience of it, most recently from the AMT Group C race engine. However, there was no advantage in terms of engine height in the car, and marketing were not keen. We would also have to find somewhere to put the oil tank. So, as we were comfortably exceeding the target power requirement, it was decided to try and make the standard sump work. The standard sump at the time had originated from the Lagonda, and was shallower than the original 16-valve V8 type. The Lagonda had a very low bonnet line, so the chassis was designed to set the engine as low as possible, which resulted in the shallow sump design. The Virage chassis design was based on the Lagonda chassis and so the engine used the same sump. Inevitably this resulted in oil surge problems at the G levels achievable in the Virage, so side flap valves were fitted to control oil surge during hard cornering. The rate of acceleration and deceleration achievable by the Vantage created another problem of fore and aft oil surge. This required additional baffles and closing plates to avoid oil starvation to the bearings, etc. Considerable hard circuit testing with Mr Jones at the helm gave confidence that the baffled sump would maintain the oil supply to the bearings. The available rate of acceleration also pushed the oil issuing from the camshafts, etc. firmly to the back of the cylinder heads, making the new rear oil drain back to the crankcase an absolute must. A modern, full synthetic engine oil was considered essential to such high engine performance and accordingly Mobil 1 was used for all development test work and approved for the Vantage.

Exhaust Gas and Noise Emissions

R egardless of the low production volumes, all Aston
Martin production cars are fully certified to comply with current legislative requirements. The Vantage
would be no exception. As anticipated, base engine emissions were low, particularly Nox, so exhaust gas emissions
were never a problem. The engine was designed for world
market compliance, so European emission levels were relatively easy to meet. If we had followed our original world
market intention, we could have sold to the USA up to
1996 before we needed to comply with the on-board
diagnostic (OBD) requirement *(inclusive of the 12-month
dispensation for existing production models)*. OBD compliance
would have then required a massive investment, which
would have been difficult to recoup at our low production
volume. But think what fun the Americans could have
had up to 1996 – they would have loved it!

Initially we had certified the car to the 77 dB(A) drive-
by noise limit. But for 1997 we had to comply with the
new 74 dB(A) regulations, which resulted in a more re-
strictive exhaust system and the higher 3.53:1 axle ratio.
The latter is primarily to reduce engine rpm during the
test. Some power loss was inevitable due to the increased
exhaust backpressure. My old friend Mike Marriott
showed me some bypass valves that could help to get
around the backpressure problem. But they would need
to be operated by a signal from the engine management
system. This would involve some input from Steve
Armitstead on the electronics. But by this time Steve was
up to his neck in work on the electronics for the new V12
engine, which took priority. So we had to put up with
the increased backpressure. To counter this, the engine

was redeveloped with a higher 2:1 supercharger drive ratio, still within the design speed limitations for the supercharger. Power was recovered back to 550 bhp but torque has increased to more than 575 lb/ft, which more than offset any effect that the higher axle ratio might have on acceleration times – 100 mph still being available in 10.1 seconds from stationary. The higher axle ratio also gave a true 200 mph potential in fifth gear. Not bad for a four-seat, two-ton-plus motor car.

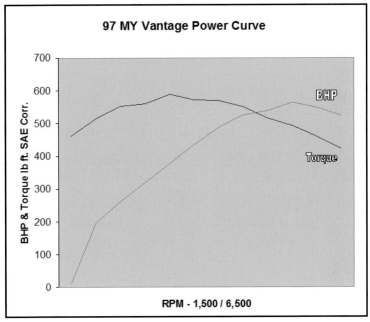

Figure 20.1 Bit of a drop in bhp from the 585 bhp of the first Vantage engine, but look at the torque!

Fuel consumption was a very high priority at the outset and a great deal of importance was attached to this during the design. We tested the car with several different drivers over a set route of mixed traffic conditions. The lowest result was 17 mpg and the highest was 20 mpg. The 20 mpg was achieved by the guy with the biggest boots, Bev Jones, and he did this a couple of times. So it was no fluke, whereas the 17 mpg was a one-off. Twenty mpg in real-world driving from a car weighing two tons with 550 bhp on tap has to be pretty remarkable, but nobody notices anymore. Nevertheless the original theory would seem to have been born out: supercharging can provide smooth unfussy power over a very wide engine range with low exhaust emissions and good fuel consumption. For comparison, the 1980 development to reduce fuel consumption tested over the same route gave 16.92 mpg. This was on the leaded four-star fuelled, non-catalysed Oscar India V8 Coupe with 305 bhp available, although to be fair, this had automatic transmission. The 305 bhp may seem low but the 1980 development had prioritised a very strong torque curve over top end bhp. It made for a very driveable car, one of my favourites.

The average fuel consumption for the Vantage over the total mileage of the above test, involving various drivers, worked out at 19 mpg. Official test figures gave a combined figure of 13.1 mpg.

200 mph

It is surprising how attitudes have changed since we developed the Vantage some years ago. There was a definite concern at the time, that we should not overemphasise the engine output due to concerns about the social acceptability of such a device. The quoted 550 bhp was a very conservative figure for the engine for this reason. The original 1992 engine produced 436.3 Kw or 585 bhp at 6,208 rpm with 745.8 Nm or 550 lb/ft at 4,058 rpm. The same applied to any attempt at 200 mph, this was raised during high-speed testing but dismissed as not being socially acceptable at the time. In fact the whole of the maximum speed development programme was axed from my original development programme. So we had to put a figure of 197.6 mph for the car, which the German TUV had accepted. Later there was a change of attitude as more manufacturers joined the 200 mph club.

Luckily this change of heart happened just before we were about to deliver the first cars to their owners. Someone decided that it would be a good idea to allow a motoring journalist to take a car to NARDO to do a max speed test. I was against this and wanted to instrument the car and do the first test to make sure everything was OK before letting the journalist loose in the car. I was overruled on this but was allowed to have Bev Jones drive the car with the journalist, with a technician, Roger Date, and spares as backup. I prepared a list of things to do and they duly set off with the car. The next that I heard from Bev Jones was that the car had failed. It had got up to near max speed and then hesitated and plumed smoke out of the exhaust. We got the car back and took out the engine, rebuilt it and ran it on the test bed and all seemed

OK. It had scuffed a piston badly. This time when we rebuilt it I had put some thermo tabs on the engine to identify any irregularities.

The car went back and the same thing happened again, by this time I was getting some very odd looks. When we got the car back and looked at the thermo tabs the one on the inlet manifold showed that it had gone up to 90 degrees Celsius – this was on the outside of the manifold, so you can imagine how hot the air temp was inside! This showed that the intercoolers were not working. We knew that the ones on the engine were OK, so we took the air dam down. What we found was that the air tunnels that we had designed into the inner air dam during our inter-cooler development had been removed, there was no way to direct the air flow through the intercooler radiators at all. Also the gauze stone guards in the entry points had no flow through them at speed.

I tried to find out who had done this but all went very quiet, it was not what I had signed off. I can remember that when Bev and I were designing the air dam we had some problems. Someone (not to be named) kept changing our design by instructing the guy that was making the prototype to simplify it by leaving the ducting out. It was not until I had had words that it stopped. I can only imagine that it happened during the three-team period, despite my daily vigil. But I had my suspicions. During this time I had made two trips up the line each day, one in the morning and one in the afternoon, just to check what was going on. Anyway we rebuilt the air dam to the same spec as the development cars and re-specified a flow-tested gauze and the car was ready. But I think that everyone had given up on the idea, so we didn't do another run. The service department reworked customer cars before they were delivered so that we didn't have customer cars running around waiting to fail.

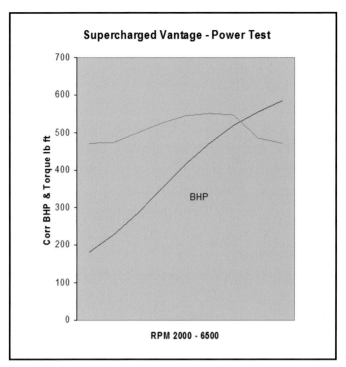

Figure 21.1 The 1992 spec Supercharged Vantage power curve (Weber engine management).

Further Development

There has been very little development of the Vantage since the original was released in 1992 other than the 1997 model year drive by noise compliance and the 600 bhp upgrade for the AML service department. We did design a system for a pair of additional injectors fitted into the intercooler to manifold plenum pipes. I had wanted to develop a water injection system for the Le Mans Vantage, but they could have equally been used for either water injection or for additional power injectors to overcome the 600 bhp limitation of the standard fuel injection system. We also did a single M112 supercharger version and an outline design for a short stroke six litre with two M112 superchargers, not to mention a compound turbo and supercharger system. Sadly none were pursued. But perhaps this was just as well, because unless we could get some weight off of the car, the driveline would be in real trouble. We were very close to the immovable object and irresistible force situation as it was.

We looked into some very interesting ideas during the development of the 32-valve engine, mostly in preparation for what could be future legislation. These ideas included:

- A heat storage battery to release stored heat from the cooling system during the next cold start, to reduce cold start emissions.

- Exhaust gas ignition to re light the exhaust gas as it entered the catalyst. To speed up catalyst light off in a position that allowed for a decent exhaust manifold and positions the catalyst away from the engine under the car to reduce under bonnet heat.

- Reed valves within the induction system to avoid backflow within a common intake system and to improve idle stability and response. A possible alternative to individual port throttles.

- Variable valve timing for reduced valve overlap at low engine speed for reduced emissions.

Figure 22.1 Single supercharged development engine.

The V8 Coupe engine, which was partly developed as the basis for the Vantage, was released in 1996, very much as it was during the Vantage development, with little in the way of further refinement. Nevertheless it produces a good balance of power and torque with 350 bhp *(11 bhp/litre/1,000 rpm)* and 368 lb/ft torque, the emphasis being on a good spread of torque as being appropriate to our heavy car. Our performance tests gave 5.8 seconds to 60 mph and 13.5 seconds to 100 mph, which was pretty brisk by any standard of the time, certainly for a heavy four-seat car with a strong but relatively long in the tooth auto transmission.

Figure 22.2 Single supercharged development engine.

But idle refinement could perhaps have been better if we had been given the budget to pursue the individual throttle route that had shown such promise.

Using the hit the throttle a split second before releasing the brake technique, standing starts can be a bit tricky with the V8 Coupe as due to some product rationalisation by the axle manufacturer the lock up ratio has been reduced. This can lead to one-side wheel spin that needs to be controlled, which could explain the difference between the magazine test results of the time and our in-house figures.

At the time of the V8 *(at one time it was referred to as the Virage S)*, we as a company were going through what can only be described as a period of indecision, certainly as far as Newport Pagnell was concerned. We had no clear plan as to how, or if, we would revitalise the Virage, which by then was becoming a bit long in the tooth. Before he left, Mike Morton had put forward

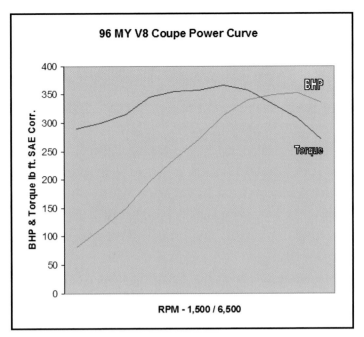

Figure 22.3 96 V8 Coupe Power Sheet 2.

several proposals from engineering, to provide a successor to the Virage, but none had been approved.

Ian Calnan, who was production manager at the time, was becoming increasingly concerned for his workforce and wanted something new to boost sales. He kept on nagging at me to see what we could give him in the way of something quicker. As things stood, this would have to be with no agreed budget or project approval. Within engineering, we were also becoming a little impatient so the engine project was started without approval as a Saturday club project, in other words 'do it in your own time'. Eventually everybody came on board and the project gained some very limited approval, although we were not allowed to do the manual transmission version, which would have been very quick. Under these circumstances, our plans to run solid lifters, a new cam profile No. 434062 which had already been designed, individual throttles and any thoughts of a stretch to six litres had to be abandoned.

I think that it was at that time that it became totally obvious that the V8 engine did not feature in the long-term plans for the new company and that what we were allowed to do for the V8 Coupe would be controlled accordingly. It was very lucky that we had done most of the Vantage before the changes started to take effect, although as I have said, we very nearly didn't put it into production. But it is a good job that we did as it bridged what would have been an awfully long wait for the new-generation Aston Martin to materialise. And we had the satisfaction that it allowed the V8 engine to go out on a high.

I suppose that there is no denying that the AML V8 engine was a very costly unit to manufacture. Internal costs alone were about £11,124/engine in 1992 with approximately 21 per cent of that figure being for labour. Most of the cost is down to every component in the engine being unique to AML and the low volumes involved (one or two per week). Other than that, it is not dirty or heavy and it packages well. Nowadays it is – at 5.3 litres – a modest capacity engine when compared with most of its contemporaries. How times have changed.

Assuming the normal requirement of an adequate supply of oil and cooling water at the right temperature and pressure, the 32-valve AML V8 engine is a very robust unit, in keeping with its predecessor, the 16-valve V8. We have had to redesign the left-side chain tensioner and modified the piston and ring pack to improve oil consumption during the production life of the engine. But it now has no known weaknesses and should provide long and trouble-free service if looked after properly.

Footnote

There is a romantic notion that the current AML V8 is a 30-year-old design – it isn't. There is no way that an engine of that age can be expected to double its power and torque output and comply with modern emission regulations without a major redesign. But it makes good reading and rightly maintains the connection back to the creator of the original AM V8, Tadek Marek, who is still held with the greatest respect and affection by all who were privileged to know him at AML. So it is not an unworthy intention, but one that can also tend to obscure the true status of the current 32-valve engine. In truth, the only parts that are the same as the original V8 are the crank nose sprockets and the woodruff key.

Notwithstanding the fundamental design differences involved in a 32-valve V8 engine as compared to the 16-valve hemi headed original, the current V8 engine is now very different. Released in 1989 as a modern 4 vpc V8 engine, it employed state-of-the-art features associated with producing environmentally clean power. When the Vantage version was released in 1992, it was the most powerful road legal production car engine in the world combined with the potential for world market compliance. And it was still the most powerful production Aston Martin engine for almost 20 years.

Last Days

Figure 23.1 Inscription stamped onto the cylinder block of the last engine (71008). The 7580 refers to the block number stamped onto the left-side bell housing flange at the rear of the block. The left cylinder head fitted to this engine is No. 7542 and the right is No. 7541.

To commemorate the last engine built at Newport Pagnell, an oil sample was taken from the sump as the engine was removed from the test bed. Mike Frost, the technical manager at Mobil Oil, made arrangement for this sample to be encapsulated and mounted as a memento. The Mobil Oil Company generously presented the resultant five numbered editions of this to AML. I was presented with the first of the five by AML.

Engine Oil Sample.

10/07/00

Origin:
Mobil Oil Company Ltd.

Usage/History:
This running in oil sample was taken from Vantage engine number 71008, which was built and tested by Chris Bennett. Vantage engine No. 71008 was the last V8 engine manufactured at Newport Pagnell, which concludes a long history of Aston Martin and Lagonda engine manufacture, at this site.

The sample was taken at 11.05 am. on Monday the 10th of July. The engine had just been removed from test cell 1, having completing its run in and power test on the preceding Saturday morning.

Verification:
Mike Frost.
Technical Manager – Mobil Oil Company.

Chris Bennett.
AML Engine Builder.

Arthur G Wilson.
V8 Power-train
Engineering Manager (Chief Engineer V8 Engines).

Figure 23.2 The last engine fresh off of the test bed with engine builders Terry Durston, Roy Robarts, Chris Bennett and Ron Russell.

Figure 23.3 Vantage Engine No 71008, the last Aston Martin engine to be built at Newport Pagnell.

Sadly Ron Russell, who was a long-term friend of mine (we had worked together in the road test and rectification department in the 1960s), passed away shortly after Figure 23.2 was taken. Ron was one of those lovely unassuming guys who was a quiet master of his craft. He was a happy and jovial friend who didn't seek the limelight but who was always ready to offer a helping hand. A sad loss to us all.

A Photo Record of the Last Days of
Engine Manufacture at Newport Pagnell

T his is a photographic record of the last days of engine manufacture at Newport Pagnell. It is no more, so it is a record of the guys that machined and built the engines that powered the exciting and beautiful cars that were built there.

Figure 24.1 The machine shop looking from the inspection bay. Cylinder blocks, sumps, etc. awaiting inspection.

Figure 24.2 Geoff Bright preparing to machine the last batch of block and head castings.

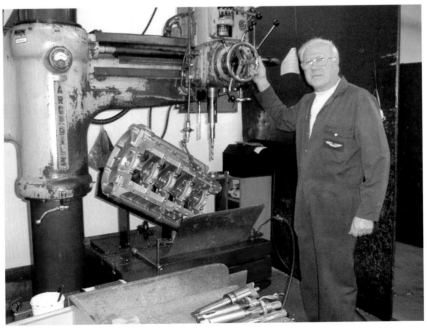

Figure 24.3 Sid Miller machining the oil pressure relief valve seat.

Figure 24.4 Des Lovett machining the main bearing bore.

Figure 24.5 Checking with the Solex gauge.

Figure 24.6 And then with the checking bar for alignment.

Figure 24.7 Cylinder block with main bearing checking bar in place.

Figure 24.8 Cylinder head camshaft boring.

Figure 24.9 Sid Miller machining the cam bore on the cylinder head.

Figure 24.10 Cutting the exhaust valve seats and valve guides.

Figure 24.11 Sid cutting the valve seats and machining the valve guide bores. My apologies to Dave Noaks, whose patience was much appreciated in getting this operation right and who usually did this job. But shift work prevented me from getting a photo of him on the job, so to speak, before things came to an end.

165

Figure 24.12 Geoff Hoden inspecting a finished machined cylinder block.

Figure 24.13 The last days for the Aston Martin engine build shop, only one engine block in a stand.

Figure 24.14 Terry Durston with 71007, his last engine. Roy Robart's last engine was 71006.

Figure 24.15 Roy Robarts prepares a freshly machined cylinder block.

Figure 24.16 Just out of the hot wash ready for liners, etc.

Figure 24.17 Cylinder liners ready to fit.

Figure 24.18 Chris Bennett fitting the cylinder liners.

Figure 24.19 Liners fitted.

Figure 24.20 Crankshaft balance machine.

Figure 24.21 Roy Robarts balancing the crankshaft.

Figure 24.22 Cleaned and ready to fit.

Figure 24.23 Conn rod set, swing weight balancing.

171

Figure 24.24 Balanced and ready to assemble.

Figure 24.25 Roy grading the main bearing clearances to 0.0011–0.0018".

Figure 24.26 Bearing shells selected, Chris and Roy carefully put the crankshaft in position.

Figure 24.27 Crank in position.

Figure 24.28 Forged Main bearing caps ready to fit.

Figure 24.29 Main bearings, rods and pistons installed.

Figure 24.30 The flat crown identifies this as a standard piston and rod assembly. Vantage has a bowled piston crown.

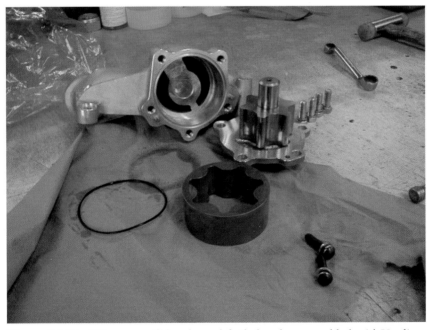

Figure 24.31 Oil pump will be cleaned fettled and re assembled with Vaseline to ensure rapid priming.

Figure 24.32 Timing chains resting in place.

Figure 24.33 A left-hand cylinder head ready to fit valves and springs.

Figure 24.34 A valve and spring assembly posing for the camera.

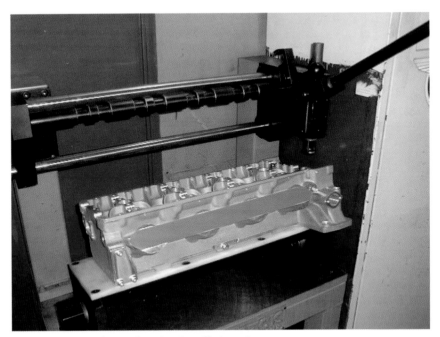

Figure 24.35 Valve and spring installation rig.

177

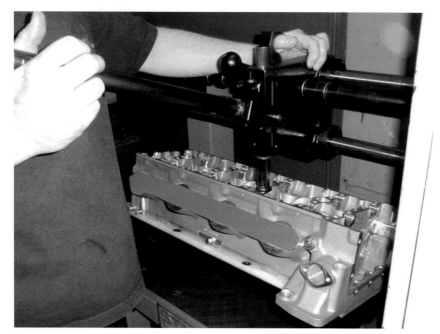

Figure 24.36 Valves and springs being installed.

Figure 24.37 Head gasket ready for the head to be fitted to normally aspirated engine.

Figure 24.38 DTIs in place ready for timing.

Figure 24.39 Both heads timed up on a normally aspirated engine.

Figure 24.40 Entrance to beds 1 and 2, past the development bed to the left behind the blue door. A V12 engine can just be seen outside of bed 2, just before the door to the pump room.

Figure 24.41 A normally aspirated engine on bed 1 prepared for test. Note the non-tuned exhaust manifolds.

Figure 24.42 The same normally aspirated engine ready for run in and power test.

Figure 24.43 Chris Bennett at the controls.

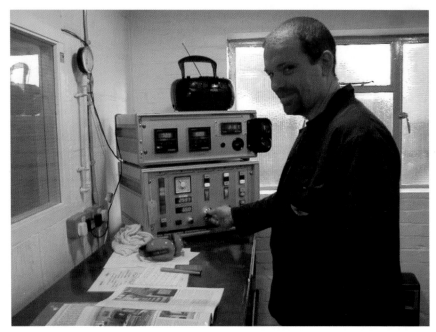

Figure 24.44 Chris Bennett running in his last engine, the Vantage engine No. 71008.